LAW, CRIME AND LAW ENFORCEMENT

TASERS AND CONDUCTED ENERGY DEVICES

SAFETY AND USE BY LAW ENFORCEMENT

Law, Crime and Law Enforcement

Additional books in this series can be found on Nova's website under the Series tab.

Additional e-books in this series can be found on Nova's website under the e-books tab.

LAW, CRIME AND LAW ENFORCEMENT

TASERS AND CONDUCTED ENERGY DEVICES

SAFETY AND USE BY LAW ENFORCEMENT

LEROY R. ELLIOTT
AND
ALEX W. RILEY
EDITORS

nova publishers
New York

For permission to use material from this book please contact us:
Telephone 631-231-7269; Fax 631-231-8175
Web Site: http://www.novapublishers.com

Additional color graphics may be available in the e-book version of this book.

Library of Congress Cataloging-in-Publication Data

ISBN 978-1-62257-463-6

Published by Nova Science Publishers, Inc. + New York

CONTENTS

PREFACE

Today, more than 12,000 law enforcement agencies in the United States use conducted energy devices (CEDs) as an alternative to conventional physical control tactics or other means of subdual. A National Institute of Justice (NIJ)-sponsored expert panel, convened to evaluate the safety and effectiveness of CEDs and concluded that law enforcement officers need not refrain from using CEDs to place uncooperative and combative subjects in custody provided that the CEDs are used in accordance with accepted national guidelines and an appropriate use-of-force policy. The panel also concluded that field use of CEDs is safe in the vast majority of cases and creates less risk of injury to officers and suspects alike, than other options of subduing uncooperative persons. This book examines the safety and use by law enforcement officers of force, tasers and other less lethal weapons with a focus on a study also of deaths following electro muscular disruption.

Chapter 1 – In its final report, an expert panel of medical professionals concludes that the use of conducted energy devices by police officers on healthy adults does not present a high risk of death or serious injury.

Chapter 2 – Law enforcement agencies continue to seek alternatives to lethal force and better methods to subdue individuals in order to minimize injuries and death. Less-lethal technologies have been used by law enforcement for this purpose extensively since the early 1990s. As of spring 2010, conducted energy devices (CEDs) causing electro muscular disruption have been procured by more than 12,000 law enforcement agencies in the United States. Approximately 260,000 CEDs have been issued to law enforcement officers nationwide. Police adoption has been driven by two major beliefs: first, that CEDs effectively facilitate arrests when suspects actively resist law enforcement; second, that CEDs represent a safer alternative than other use-of-force methods. Studies by law enforcement agencies deploying CEDs have shown reduced injuries to both officers and suspects in use-of-force encounters and reduced use of deadly force. More recently, independent researchers have come to similar conclusions, when appropriate deployment and training policies are in place.

Chapter 3 - This study looked at injuries that occur to law enforce-ment officers and citizens during use-of-force events. Most applications of force are minimal, with officers using their hands, arms or bodies to push or pull against a suspect to gain control. Officers are also trained to use various other force techniques and weapons to overcome resistance. These include less-lethal weapons such as pepper spray, batons or conducted energy devices (CEDs) such as Tasers. They can also use firearms to defend themselves or others against threats of death or serious bodily injuries.

In: Tasers and Conducted Energy Devices ISBN 978-1-62257-463-6
Editors: L. R. Elliott and A. W. Riley

Chapter 1

FINAL FINDINGS FROM THE EXPERT PANEL ON THE SAFETY OF CONDUCTED ENERGY DEVICES*

Brian Higgins

In its final report, an expert panel of medical professionals concludes that the use of conducted energy devices by police officers on healthy adults does not present a high risk of death or serious injury.

Today, more than 12,000 law enforcement agencies in the United States use conducted energy devices (CEDs) as an alternative to conventional physical control tactics or other means of subdual. An NIJ-sponsored expert panel, convened to evaluate the safety and effectiveness of CEDs, issued its final report in May 2011. The panel concluded that law enforcement officers need not refrain from using CEDs to place uncooperative and combative subjects in custody provided that the CEDs are used in accordance with accepted national guidelines and an appropriate use-of-force policy. In its report, the panel concluded that field use of CEDs is safe in the vast majority of cases and creates less risk of injury — to officers and suspects alike — than other options of subduing uncooperative persons.

In addition to investigating the effects of CEDs, the panel issued recommendations for their use. Among these were to apply CEDs for no longer than 15 seconds at a time and to limit the number of discharges to the

* This is an edited, reformatted and augmented version of the NIJ Journal/Issue No. 268, National Institute of Justice, dated October 2011.

fewest needed to control the suspect. The panel also said that, regardless of how long the CED exposure lasts, some form of medical screening and ongoing observation of individuals exposed to CEDs is crucial. Screening should start at the scene and individuals should continue to be monitored in custody for abnormal physical and behavioral changes.

CEDs, such as Tasers, generate 50,000 volts of electricity. The electricity stuns and temporarily incapacitates people by causing involuntary muscle contractions. This makes people easier to arrest or subdue. Widespread police adoption of CEDs has been driven by two major beliefs: that CEDs facilitate arrests when suspects actively resist and that they are safer than other use-of-force options. Independent researchers studying law enforcement agencies that deploy CEDs have concluded that, when used appropriately by properly trained officers, CEDs have reduced injuries to officers and suspects in use-offorce encounters and reduced use of deadly force.[1]

Nonetheless, a number of individuals have died after exposure to a CED. Some were healthy adults; many were chemically intoxicated or had some underlying medical condition. These deaths have caused law enforcement personnel and the public to ask questions about the safety of CEDs.

To answer these questions, NIJ, in cooperation with the College of American Pathologists, the Centers for Disease Control and Prevention, and the National Association of Medical Examiners, conducted a study to address whether CEDs can contribute to or be the primary cause of death and, if so, how.

To support the study, an expert medical panel was formed, composed of forensic pathologists, medical examiners and specialists in cardiology, emergency medicine, epidemiology and toxicology. The panel reviewed 300 subdual cases in which a CED was used and later the person died.

> The panel concluded that, in general, the stress of receiving a CED discharge is comparable to the stress from other-wise being physically restrained or subdued.

In the vast majority of these cases, the original medicolegal investigations concluded that CED exposure was not the cause of death. The panel conducted in-depth reviews of 22 of those 300 cases and reviewed approximately 175 peer-reviewed articles on the physiological effects of CEDs. The panel's report provides findings concerning death investigation, CED use, CED-related health effects and medical response to the use of CEDs. The panel determined that there is no conclusive medical evidence in the current body of research literature that indicates a high risk of serious injury or death to humans from the direct or indirect cardiovascular or metabolic effects of short-term CED exposure in healthy, non-stressed, non-intoxicated persons.

Field experience with CED use indicates that short-term exposure is safe in the vast majority of cases. According to the final report, the risk of death in a CED-related use-of-force incident in the general population is less than 0.25 percent (one in 400). The report notes that, based on the panel's review and confirmation of the findings of the original death investigations of 300 deaths following CED exposure, it is reasonable to conclude that CEDs do not cause or contribute to death in the large majority of cases.

The panel concluded that, in general, the stress of receiving a CED discharge is comparable to the stress from otherwise being physically restrained or subdued. Verbal altercation, physical struggle and physical restraint all generate stress that may heighten the risk of sudden death in individuals who have a pre-existing cardiac condition or certain other diseases.

Unlike the risk of secondary injury (e.g., injuries due to falling as a result of CED exposure, discussed below), the risk of death directly or primarily due to the electrical effects of CED application has not been conclusively demonstrated. The literature suggests a substantial safety margin with respect to the use of CEDs when they are used according to the manufacturer's instructions. The possibility that the effects of a CED can be directly lethal in some cases, however, cannot be excluded — though not conclusively demonstrated, plausible mechanisms of injury exist. There are anecdotal cases in which no other significant risk factor for death is known and the timing of death provides circumstantial evidence that the CED's application was the cause of death. As such, there remains at least a theoretical possibility that in rare cases, CED application could be directly or primarily responsible for death due to a confluence of unlikely circumstances.

The report states that the risk of significant injury from CEDs is also low (0.5-0.7 percent). Significant injuries associated with CED use documented in the studies reviewed by the panel included puncture wounds from CED darts (including wounds to the eye, throat and skull resulting in loss of vision, unconsciousness and seizures requiring medical care) and falls related to muscular incapacitation or intense muscle contraction.

The panel highlighted the possibility of secondary injuries resulting from the use of CEDs on tall structures or steep slopes, where individuals exposed to a CED could fall; near flammable materials (including gasoline, explosives, aerosols and propellants) that a spark from a CED could ignite; and in water, where submersion could lead to drowning. The use of CEDs also presents a risk of interfering with implantable cardiac devices, such as pacemakers, although no bad outcomes have been reported. Furthermore, the physiological effects of prolonged or repeated CED exposure are not fully understood.

The panel acknowledged that there may be circumstances in the field that require repeated or continuous exposure to a CED discharge. They emphasized that law enforcement personnel must be made aware that the associated risks are unknown and most deaths associated with CED use involved multiple or prolonged discharges. The report states that it is critical that law enforcement officers minimize or avoid multiple or prolonged activations of CEDs as a means of subduing an individual.

How CEDS Work

Most conducted-energy devices (CEDs) carried by law enforcement officers in the U.S. can operate in two modes: a drive-stun mode and a probe mode. In both modes, CEDs work by sending energy down two electrical contacts. If the contacts are touching an object, a conduit, the electricity will flow from one contact to the other through that object, closing the circuit.

An open circuit (when there is no conduit) on a CED can generate up to 50,000 volts (the peak open circuit arcing voltage). When the circuit is closed, such as when the probes are embedded in someone's torso, a CED may produce approximately 5,000 volts (the amount will depend on the model). For comparison, the standard U.S. wall outlet generates 120 volts. It is, of course, extremely dangerous to receive a shock from a wall outlet. So, how is it possible for a human body to safely receive 5,000 volts from a CED?

To answer that question, we need to look at another measure of energy: current. If we think of electricity as water flowing through a pipe, rather than electrons traveling along a wire, then voltage is the pressure it takes to push water through the pipe, while current is the rate at which the water flows. Electrical outlets have a high, continuous current — after all, we expect them to supply us with a high, steady stream of energy so our lights, appliances and electronics work without interruption.

CEDs, on the other hand, have a low, pulsed current. After the probes are attached to skin or clothing, the trigger activates a five-second series of low-current pulses. It may, for example, activate 19 low-current pulses per second that last for 30 microseconds (30 millionths of a second) each. It should be noted that some versions of CEDs in use can deliver multiple discharges if the trigger is pressed again after the first cycle or prolonged and uninterrupted discharges if the trigger is held down continuously.

CEDs will have different effects on people depending on which mode they are in and officers may use them for different purposes (incapacitation versus deterrence).

In probe mode, CEDs use compressed nitrogen to fire two barbed probes (sometimes called darts) at a target, imbedding themselves in the target's skin or clothing. Unlike in drive-stun mode, the probes are not directly next to one another and the electrical current is spread out across more tissue. When the trigger is pulled, electricity travels along thin wires attached to the probes. In addition to causing pain, the electrical current interferes with the target's neuromuscular system. The interference causes involuntary muscle contractions, temporarily incapacitating the target and making him or her easier to arrest or subdue.

In drive-stun mode, when a CED's contacts are applied directly to a target, CEDs do not have the same incapacitating effect that they usually do in probe mode. Because the electrical contacts are closer together, they do not engage or electrically excite as much tissue and, consequently, do not temporarily interfere with a person's neuromuscular system. They do, however, cause pain, which may deter an individual from continuing his or her behavior.

The report also states that the safety margins of CED use in healthy adults may not apply to everyone. The effects of CED exposure on small children, those with diseased hearts, the elderly, pregnant women and other potentially at-risk individuals are not clearly understood, and more data are needed. Law enforcement personnel should minimize or avoid use of a CED on members of these populations.

In addition to recommendations governing the use of CEDs, the panel issued advice in the event a death occurs following the use of a CED. The panel recommended that all deaths following deployment of a CED should be subject to a complete medicolegal investigation. This investigation should include an autopsy by a forensic pathologist and a medically objective investigation independent of law enforcement. In addition to the conventional information collected in a death investigation, investigators should collect information specific to the CED-related death, such as the manner in which CED darts or prongs were applied and where they were applied.

Finally, the panel recommended that law enforcement personnel maintain an ongoing dialogue with medical examiners or coroners and emergency physicians to discuss effects of all use-of-force applications, including those involving CEDs, and evaluate procedures involving life preservation, injury prevention and evidence collection.

> The report states that it is critical that law enforcement officers minimize or avoid multiple or prolonged activations of CEDs as a means of subduing an individual.

For More Information

[1] Read the final report, *Study of Deaths Following Electro Muscular Disruption,* at https://www.ncjrs.gov/pdffiles1/nij/233432.pdf.

[2] Read the NIJ Research in Brief, *Police Use of Force, Tasers and Other Less-Lethal Weapons,* at https://www.ncjrs.gov/pdffiles1/nij/232215.pdf.

[3] Visit NIJ's Web topic page (http://www.nij.gov/topics/technology/less-lethal/conducted-energy-devices.htm) and read an article in the *NIJ Journal,* "Police Use of Force: The Impact of Less-Lethal Weapons and Tactics" (http://www. nij.gov/journals/267/use-of-force.htm).

End Note

[1] Smith, Michael R., Robert J. Kaminski, Geoffrey P. Alpert, Lorie A. Fridell, John MacDonald, and Bruce Kubu, *A Multi-Method Evaluation of Police Use of Force Outcomes,* Final report to the National Institute of Justice, grant number 2005-IJCX-0056, July 2010, NCJ 231176, available at https://www.ncjrs.gov/ pdffiles1/nij/grants/231176.pdf .

In: Tasers and Conducted Energy Devices ISBN 978-1-62257-463-6
Editors: L. R. Elliott and A. W. Riley

Chapter 2

STUDY OF DEATHS FOLLOWING ELECTRO MUSCULAR DISRUPTION*

United States Department of Justice

ACKNOWLEDGMENTS

The National Institute of Justice gratefully acknowledges the following individuals. Their information, insight and knowledge benefited the development of this report.

Larry Amerson
Sheriff, Calhoun County, Ala.
Albert Arena
Project Manager
International Association of Chiefs of Police
Laura Beck
Corporal, Maryland State Police
Deborah Boelling
Assistant Director

David Hammel
Detective Sergeant (Ret), Maryland State Police
Joseph Kocab
Chief, Brooklyn Heights, Ohio, Police Department
Mark Marshall
Chief, Smithfield, Va., Police Department
James Martyn
Lieutenant, Maryland State Police

* This is an edited, reformatted and augmented version of the U.S. Department of Justice, Office of Justice Programs, NCJ 233432, dated May 2011.

St. Louis, Mo., Police Academy

John Branham
Sergeant, Maryland State Police

William L. Brewer
Captain, Birmingham, Ala..,
Police Department

Vernon Busby
Officer, Phoenix, Ariz.,
Police Department

John Cook
Detective Sergeant, Maryland
State Police

Dan Cornwell
Captain, Maryland State Police

Dawn Diedrich
Deputy Director of Legal
Services
Georgia Bureau of Investigation

Lisa Erazo
Project Coordinator
International Association of
Chiefs of Police

John Firman
Director
International Association of
Chiefs of Police Research Center

Alan Goldberg
Captain, Montgomery County,
Md., Police Department

John Grant
Senior Program Manager
International Association of
Chiefs of Police

James McMahon
Deputy Executive Director
International Association of
Chiefs of Police

Jeffrey B. Miller
Colonel, Superintendent (Ret),
Pennsylvania State Police

Peter Modafferi
Chief of Detectives, Rockland County,
N.Y., District Attorney's Office

Karen Montejo
Chief, Miami-Dade,
Fla., Police Department

Dan Rosenblatt
Executive Director
International Association of
Chiefs of Police

Michael A. Spochart
Lieutenant, U.S. Capitol Police

Sabrina Tapp-Harper
Lieutenant, Baltimore,
Md., Police Department

Douglas Ventre
Lieutenant, Cincinnati, Ohio, Police
Department

Otis Whitaker
Sergeant, Maryland State Police

Ray Wojcik
Lieutenant (Ret),
Maryland State Police

EXECUTIVE SUMMARY

Law enforcement agencies continue to seek alternatives to lethal force and better methods to subdue individuals in order to minimize injuries and death. Less-lethal technologies have been used by law enforcement for this purpose extensively since the early 1990s. As of spring 2010, conducted energy devices (CEDs) causing electro muscular disruption have been procured by more than 12,000 law enforcement agencies in the United States. Approximately 260,000 CEDs have been issued to law enforcement officers nationwide. Police adoption has been driven by two major beliefs: first, that CEDs effectively facilitate arrests when suspects actively resist law enforcement; second, that CEDs represent a safer alternative than other use-of-force methods. Studies by law enforcement agencies deploying CEDs have shown reduced injuries to both officers and suspects in use-of-force encounters and reduced use of deadly force. More recently, independent researchers have come to similar conclusions, when appropriate deployment and training policies are in place.

Nonetheless, a number of individuals have died after exposure to a CED during law enforcement encounters. Some were normal, healthy adults; many were chemically intoxicated or had heart disease or mental illness. These deaths have given rise to questions from both law enforcement personnel and the public regarding the safety of CEDs. Because many gaps remain in the body of knowledge with respect to the effects of CEDs, the National Institute of Justice (NIJ), the research, development and evaluation agency of the U.S. Department of Justice, conducted a study, Deaths Following Electro-Muscular Disruption, to address whether CEDs can contribute to or be the primary cause of death and, if so, by what mechanisms. The study was directed by a steering group that included NIJ, the College of American Pathologists, the Centers for Disease Control and Prevention, and the National Association of Medical Examiners.

To support the study, the steering group appointed a medical panel composed of forensic pathologist/medical examiners and other relevant physicians or specialists in cardiology, emergency medicine, epidemiology and toxicology. To avoid a conflict of interest, no panelists were chosen who had worked as litigation consultants for or against CED manufacturers. This report contains the findings and recommendations of the medical panel.

In 2008, NIJ released its interim report, *Study of Deaths Following Electro Muscular Disruption: Interim Report.* Among other findings, that report stated, "Although exposure to CED is not risk free, there is no conclusive medical evidence within the state of current research that indicates a high risk of serious injury or death from the direct effects of CED exposure." The interim report described the risks associated with the use of CEDs and provided a set of accepted research findings in its summary. The report also provided recommendations for death investigation, medical response and further research. Although this final report provides additional, significant detail to many of the findings in the interim report, the study panel's interim findings still represent its consensus on the issue of risks associated with CED use.

This final report provides findings concerning death investigation, CED use, CED-related health effects, and medical response. The panel recommends a thorough review of the entire report and the associated research literature for medicolegal personnel and those making decisions concerning CED deployment and associated policies. The following findings are provided as those of most general interest to date.

There is no conclusive medical evidence in the current body of research literature that indicates a high risk of serious injury or death to humans from the direct or indirect cardiovascular or metabolic effects of short-term CED exposure in healthy, normal, nonstressed, nonintoxicated persons. Field experience with CED use indicates that short-term exposure is safe in the vast majority of cases. The risk of death in a CED-related use-of-force incident is less than 0.25 percent, and it is reasonable to conclude that CEDs do not cause or contribute to death in the large majority of those cases.

Law enforcement need not refrain from using CEDs to place uncooperative or combative subjects in custody, provided the devices are used in accordance with accepted national guidelines and appropriate use-of-force policy. The current literature as a whole suggests that deployment of a CED has a margin of safety as great as or greater than most alternatives. Because the physiologic effects of prolonged or repeated CED exposure are not fully understood, law enforcement officers should refrain, when possible, from continuous activations of greater than 15 seconds, as few studies have reported on longer time frames.

All deaths following deployment of a CED should be subject to a complete medicolegal investigation, including a complete autopsy by a forensic pathologist in conjunction with a medically objective investigation that is independent of law enforcement. The complete investigation should include the collection of information specific to CED-related deaths, such as the manner in which and the location where CED darts or prongs were applied. A recommended checklist is contained in chapter 11, "Considerations in Death Investigation," pages 36-37 in this report.

Unlike the risk of secondary injury due to falling or puncture, the risk of human death due directly or primarily to the electrical effects of CED application has not been conclusively demonstrated. However, there are anecdotal cases where no other significant risk factor for death is known. Additionally, current research does not support a substantially increased risk of cardiac arrhythmia in field situations, even if the CED darts strike the front of the chest. There are anecdotal cases where no other significant risk factor for death is known and where the temporal association provides weak circumstantial evidence of causation. The panel reviewed studies on ventricular fibrillation with respect to dart placement, demonstration of ventricular fibrillation, pulseless ventricular tachycardia, pulseless electrical activity in animals, and anecdotal examples of capture in humans wearing cardiac pacemakers or defibrillators. These studies suggest plausible but unproven mechanisms for unusual and rare cases of death due to a confluence of unlikely circumstances.

In general, the stress of receiving CED discharge(s) should be considered to be of a magnitude that is comparable to the stress of other components of subdual. All aspects of an altercation (including verbal altercation, physical struggle or physical restraint) constitute stress that may heighten the risk of sudden death in individuals who have pre-existing cardiac or other significant disease.

Caution is urged in using multiple or prolonged activations of CED as a means to accomplish subduing the individual. There may be circumstances where repeated or continuous exposure is required; law enforcement personnel should be aware that the associated risks are unknown and that most deaths associated with CED use involve multiple or prolonged discharges.

We offer this report to the police community, the medical community and the public as a contribution to the many considerations necessarily involved in the use of CEDs and other types of force by law enforcement. We offer this report to our colleagues involved in all aspects of medicolegal death investigation to educate them on our findings and to offer possible approaches to their individual case investigations. We know full well that every case is unique and that it is extremely difficult to generalize findings or techniques. We in no way imply that our conclusions or suggestions are the only way to proceed. We offer these for consideration as aids that might be beneficial in formulating a more complete understanding of the circumstances, mechanisms or pathophysiology in determining the cause and manner of death.

It is recommended that law enforcement maintain an ongoing dialogue with medical examiners/coroners and emergency physicians to discuss effects of all use-of-force applications (CED use and other modalities) and evaluate procedures involving life preservation, injury prevention and evidence collection.

Any expert panel brings with it certain limitations. These limitations are due not only to the limitations of our knowledge but also to the perspectives that the panel members bring to the table. This is particularly true with respect to the determination of the cause and manner of death. These differences are not capricious, but derive from varying philosophical viewpoints and traditions regarding how these deaths should be placed within specific cultural and legal contexts. The conclusions in this report represent a strong underlying consensus. In instances when there were disagreements over specific classifications or diagnostic categorizations, the discussions did not reflect differences in the understanding of basic underlying scientific principles but rather the differences inherent in specific jurisdictional-related and historic practices. In fact, there was a strong consensus regarding the principles of these conclusions even in the context of differences in how they might be phrased. In addition, the report is based upon the information available to the panel at this writing. As scientific understanding advances, the opinions of panel members may change to accommodate new findings.

Findings and conclusions of the research reported here are those of the authors and do not reflect the official position and policies of their respective organizations or the U.S. Department of Justice. The products, manufacturers and organizations discussed in this document are presented for informational purposes only and do not constitute product approval or endorsement by the U.S. Department of Justice.

METHODOLOGY

This study was directed by a steering group with representation from the National Institute of Justice (NIJ), the College of American Pathologists, the Centers for Disease Control and Prevention, and the National Association of Medical Examiners. To support the study, the steering group appointed a medical panel composed of forensic pathologists/medical examiners and other relevant physicians or specialists in cardiology, emergency medicine, epidemiology and toxicology. To avoid a conflict of interest, no panelists were chosen who had worked as litigation consultants for or against conducted energy device (CED) manufacturers. This report contains the findings and recommendations of the medical panel.

In formulating the findings reported here, the panel conducted mortality reviews of CED-related deaths and reviewed the current state of medical research relative to the effects of CEDs. The panel considered nearly 300 CED-related deaths. In these incidents, (a) CED(s) was (were) deployed by (a) law enforcement officer(s) on an individual who later died. In the vast majority of these cases, the original medicolegal investigation concluded that the CED played no role in the death. The panel concentrated its review on those cases in which a CED was listed on the death certificate. NIJ and the International Association of Chiefs of Police worked with several law enforcement agencies to collect information in 22 specific, documented cases involving CED deployment and death. Time and the availability of complete case documentation (from the initial 911 call through forensic autopsy) limited the number of field-based cases reviewed and discussed by the medical panel. However, the cases reviewed were varied and considered representative of all medicolegal cases of death following CED deployment. These reviews were intended to elucidate the relationships between CED use and suspect injury and death and to assist in the development of the material in this final report. The medical panel did not make conclusions that question the findings by any official certifier of death in any specific case. Mortality reviews have included analyses of complete autopsies, findings from the scene investigation, post-exposure symptoms, post-event medical care, and especially the extent, if any, of natural disease or chemical substances in a decedent. The panel reviewed theoretical case scenarios to identify important case-related and interpreted issues regarding the cause, manner and circumstances of death. The panel also examined the currently recognized causes of sudden deaths, chiefly involving physical, cardiac, pulmonary, metabolic and thermoregulatory mechanisms.

Evaluation of mortality following the use of CEDs is often challenging because of several factors: some of the necessary case-specific information can be lacking, human research studies are limited, and the findings in animal studies may not be extrapolated to humans. There are also variations among medical examiners and coroners in the stylistic methods and choices of words used to describe the causes of death and to classify the manner of death. For a broad review such as this one of the safety of CEDs, these considerations can compromise case identification and statistical reviews of mortality following deployment of CEDs.

This report provides a consensus view of the panel members from a complete review of the available peer-reviewed research literature and extensive information concerning the use of CEDs in the field. The findings have been limited to those conclusions that can be based on current understanding of the available research and literature. A comprehensive literature search was conducted to compile and catalog peer-reviewed research articles that addressed the effects of CED on human subjects. Several resources were used to locate articles, books, news reports, websites, and other literature dealing with the use of CEDs (i.e., stun guns and other nonlethal electrical weapons), including, but not limited to: *Medline, PubMed, ScienceDirect, ProQuestJStor, Applied Science and Technology Abstracts* and *Lexis-Nexis.* More than 2,500 sources were identified, of which approximately 175 were selected for this study (i.e., peer-reviewed journal articles, which focused on the physiological effects of CED use). These selected references were divided and distributed to an external panel of forensic pathologists who reviewed and rated each article for scientific quality and relevance. These assessments were used to identify the most important research articles for consideration by the medical panel in this study. In addition, the articles are cited throughout this final report to support specific conclusions. Finally, through the National Association of Medical Examiners, the assessments are available to the medicolegal community for reference in death investigations. The panel urges continued research to improve the medical understanding of CED effects and has made specific recommendations throughout this report in that regard. Due to time constraints, some of the most recent research for this report was reviewed by panel members only.

The panel also consulted stakeholders, experts and other interested parties, such as human rights groups, law enforcement professionals, clinical physicians, research scientists and manufacturers of CEDs. The panel observed more than 30 presentations by these invited experts. It met nine times over three years to discuss these findings and debate their significance to the investigations and certifications of deaths when CEDs are involved. This report represents the panel's best efforts of collaboration and mutual respect for our many divergent points of view and perspectives.

1. Continued Use of CEDs by Law Enforcement

Conducted energy devices (CEDs) are commonly used by law enforcement agencies. Their use is associated with overall decreases in suspect and officer injuries when deployed with appropriate agency policies.[1] However, exposure to CED is not risk-free. The safety of these weapons has been the subject of controversy. CED deployment has been associated with in-custody sudden deaths. Comprehensive, independent studies have examined the experience of police agencies with respect to the decision to deploy CEDs. These studies indicate that CED deployment by an agency decreases the likelihood of injuries to suspects and officers.[1-3] Field experience with CED use indicates that exposure is safe in the vast majority of cases.[4-6] One prospective study observed a 0.25 percent risk of serious injury (head trauma or rhabdomyolysis) with CED use, much less than that observed for other subdual options.[6] Other studies also indicate that CED-related injuries and deaths are uncommon, especially in comparison to other force options.[7] One review showed that officer and subject injury rates were much lower during CED use compared to use of empty-handed physical skills, incapacitating spray or batons, while another indicated that injury rates were substantially lower with the use of incapacitating sprays and CEDs.[1,8]

It should be noted that arrestees who are involved in use-of-force incidents are by nature at higher risk for serious complication and death relative to the overall population. These individuals are more likely to be drug-intoxicated, be mentally ill or have serious underlying medical conditions.[6] There are more than 600 arrest-related deaths in the United States each year and roughly 1 million incidents in which police use or threaten to use force.[9,10] Nonetheless, the CED is cited as a causative or contributory factor in very few arrest-related deaths each year.[9] In this context, the relative risk of CED deployments appears to be lower than other use-of-force options.

There is no conclusive medical evidence within the state of current research that indicates a high risk of serious injury or death from the direct or indirect cardiovascular or metabolic effects of short-term CED exposure in healthy, normal, nonstressed, nonintoxicated persons.[11] Current medical research in humans and animals suggests that a single exposure of less than 15 seconds from a TASER® X-26™ or similar model CED is not a stress of a magnitude that separates it from the other stress-inducing components of restraint or subdual.[12] Based on cases reviewed by this panel, most adverse reactions and deaths associated with CED deployment appear to be associated with multiple or prolonged discharges of the weapons. There is limited research with regard to exposures of greater than 15 seconds.[13,14] Further, extended CED exposure may not be effective in the subdual of some individuals with high levels of drug intoxication or mental illness. Therefore, if the CED is ineffective in subduing an individual after a prolonged exposure, law enforcement officers should consider other options.

Conclusions and Recommendations

From a purely medical perspective, law enforcement need not refrain from deploying CEDs to place uncooperative or combative subjects in custody, provided the devices are used in accordance with accepted national guidelines and appropriate use-of-force policy.[15,16] Ideally, use-of-force policy development and post-incident review should be done in consultation with forensic and/or medical experts.

REFERENCES

[4] MacDonald, JM; Kaminski, RJ; Smith, MR. The effect of less-lethal weapons on injuries in police use-of-force events. *Amer J Pub Health*. 2009, 99, 1-7.

[5] Smith, MR; Kaminski, RJ; Alpert, GP; et al. A multi-method evaluation of police use of force outcomes: final report to the National Institute of Justice. Columbia, SC: University of South Carolina, 2009.

[6] Taylor, B; Woods, D; Kubu, B; et al. *Comparing safety outcomes in law enforcement agencies that have deployed conducted energy devices and a matched comparison group that have not: a quasi-experimental evaluation*. Washington, DC: Police Executive Research Forum. 2009.

[7] Eastman, AL; Metzger, JC; Pepe, PE; et al. Conductive electrical devices: A prospective, population-based study of the medical safety of law enforcement use. *J Trauma: Inj Infect Crit Care*. 2008, 64, 1567-1572.

[8] Angelidis, M; Basta, A; Walsh, M; et al. Injuries associated with law enforcement use of conducted electrical weapons. *Acad Emer Med*. 2009, 16, S229.

[9] Bozeman, WP; Hauda, WE; Heck, JJ; et al. Safety and injury profile of conducted electrical weapons used by law enforcement officers against criminal suspects. *Ann Emer Med*. 2008, 20, 1-10.

[10] Ho, JD; Heegaard, WG; Dawes, DM; et al. Unexpected arrest-related deaths in America: 12 months of open source surveillance. *West J Emer Med*. 2009, 10, 68-73.

[11] Jenkinson, E; Neeson, C; Bleetman, A. The relative risk of police use-of-force options: evaluating the potential for deployment of electronic weaponry. *J Clin Forensic Med.* 2006, 13, 229-241.

[12] Mumola, CJ. *Arrest-related deaths in the United States, 2003-2005.* Washington, DC: U.S. Department of Justice, Bureau of Justice Statistics. 2007.

[13] Durose, MR; Langa, PA; Smith, EL. *Contacts between police and the public, 2005.* Washington, DC: U.S. Department of Justice, Bureau of Justice Statistics. 2007.

[14] Bozeman, WP; Barnes Jr, DG; Winslow III, JE; et al. Immediate cardiovascular effects of the Taser X26 conducted electrical weapon. *Emer Med J.* 2009, 26, 567-570.

[15] Ho, JD; Dawes, DM; Cole, JB; et al. Lactate and pH evaluation in exhausted humans with prolonged TASER X26 exposure or continued exertion. *Forensic Sci Int.* 2009, 190, 80-86.

[16] Jauchem, JR; Seaman, RL; Klages, CM. Physiological effects of Taser C2 conducted energy weapon. *Forensic Sci Med Pathol.* 2009, 5, 189-198.

[17] Ho, JD; Dawes, DM; Cole, JB; et al. *Human physiological effects of a civilian conducted electrical weapon application.* Minneapolis, MN: Hennepin County Medical Center Study. 2008.

[18] American Medical Association. *Use of Tasers by law enforcement agencies, report 6 (A-09) of the Council of Science and Public Health.* Washington, DC: American Medical Association. 2009.

[19] International Association of Chiefs of Police. *Electronic control weapons, a model policy of the International Association of Chiefs of Police.* Alexandria, VA: International Association of Chiefs of Police, 2007.

2. Potential for Moderate, Severe or Secondary Injury

The question often arises whether injuries result from CED exposure, and, if so, to what degree of severity. Answers to these questions are important for several reasons. First, the public and law enforcement agencies need to know the risks of injury in order to have a realistic understanding of risks to persons subjected to CED exposure. This will allow police agencies to develop protocols that minimize the risk of injury and will help the public place CED-related injury in the proper context when CEDs are used by law enforcement personnel. Medical examiners, coroners, other investigators and emergency medical personnel need to understand the types of injuries that can be expected as well as their frequency so they can adequately investigate or treat injuries resulting from CED exposure.

Information to address these questions has been derived from case reports of documented CED-related injuries in humans and from descriptive studies, both prospective and retrospective, of injuries observed in populations following CED exposure.[1-12] Also, some potential injuries have been identified through review of unpublished case reports.

A practical definition of moderate and severe CED-related injury has been published.[2] Moderate injury requires inpatient treatment and/or is expected to result in no more than a moderate long-term disability. Severe injury involves a threat to life or requires inpatient treatment and is expected to result in severe long-term disability. The potential for moderate or severe injury related to CED exposure is low.[2,5,7,9,10,13-16] Based on published studies, significant injury has been noted in less than 0.5 percent of those experiencing a CED deployment, and has been estimated not to exceed 0.7 percent.[1] However, darts may cause puncture wounds or burns.[9] Puncture wounds to an eye from a dart could lead to loss of vision.[4,6] Pharyngeal (throat) perforation by a dart has also been reported.[11] Potentially fatal head injuries or skeletal fractures may result from falls due to muscle incapacitation or intense muscle contraction.[8,10] CED strikes to the head have resulted in dart penetration of the skull, and in unconsciousness and seizures requiring medical care.[3,10] CEDs can potentially produce other secondary or indirect effects that may result in death. Examples include:

1. Using a CED against a person on a steep slope or on a tall structure, resulting in a fall with traumatic injuries.
2. Ignition risk due to sparks from a CED used near flammable materials such as gasoline, explosives, volatile inhalants such as aerosol sprays, or the flammable propellant used in pepper spray.
3. Using a CED on a person who is in water, resulting in submersion or drowning.

Conclusions and Recommendations

In summary, the risk of moderate or severe injury or death from a CED exposure, whether the injury is directly due to darts or indirectly due to secondary events (falls, fractures, etc.) is probably less than 1 percent. Evidence from use in the field has shown that the risk of death in a CED-related incident is ≤ 0.25 percent.[2] These studies do not conclude that all the deaths were attributable to CED use. The panel views this as an acceptable level of risk when potential benefits of CED use are considered, such as reductions of serious injuries to suspects and law enforcement officers and the risk associated with other lethal and less-lethal options, when used in accordance with appropriate agency policies.[17,18] Further study is needed to better characterize the scope and severity of direct and indirect injuries caused by CED use.

REFERENCES

[1] Bozeman, WP; Winslow, JE. Medical aspects of less lethal weapons. *Internet J Rescue Disaster Med.* 2005, 5, 1-11. Available from: ISPUB.com, Sugar Land, TX. Accessed June 14, 2010.

[2] Bozeman, WP; Hauda II, WE; Heck, JJ; et al. Safety and injury profile of conducted electrical weapons used by law enforcement officers against criminal suspects. *Ann Emer Med.* 2009, 53, 480-489.

[3] Rehman, TU; Yonas, H; Marinaro, J. Intracranial penetration of a TASER dart. *Amer J Emer Med.* 2007 25, 733, e3-e4.

[4] Chen, SL; Richard, CK; Murthy, RC; et al. Perforating ocular injury by Taser. *Clin Exper Ophth.* 2006, 34, 378-380.

[5] Ordog, GJ; Wasserberger, J; Schlater, T; et al. Electronic gun (Taser) injuries. *Ann Emer Med.* 1987, 16, 73-78.

[6] Han, JS; Chopra, A; Carr, D. Ophthalmic injuries from a TASER. *J Can Assoc Emer Physicians.* 2009, 11, 90-93.

[7] Smith, MR; Kaminski, RJ; Rojek, J; et al. The impact of conducted energy devices and other types of force and resistance on officer and suspect injuries. *Policing: Int J Police Strategies Manage.* 2007, 30, 423-446.

[8] Sloane, CM; Chan, TC; Vilke, GM. Thoracic spine compression fracture after TASER activation. *J Emer Med.* 2008, 34, 283-285.

[9] Nanthakumar, K; Billingsley, IM; Masse, S; et al. Cardiac electrophysiological consequences of neuromuscular incapacitating device discharges. *J Amer Coll Cardiol.* 2006, 48, 798-804.

[10] Mangus, BE; Shen, LY; Helmer, SD; et al. Taser and Taser associated injuries: A case series. *Amer Surgeon.* 2008, 74, 862-865.

[11] Al-Jarabah, M; Coulston, J; Hewin, D. Pharyngeal perforation secondary to electrical shock from a Taser gun. *Emer Med J.* 2008, 25, 378.

[12] Rivera-Garcia, LE; Crown, LA; Smith, RB. Overview of electronic weapon injury and emergency department management. *Amer J Clin Med.* 2008, 5, 46-49.
[13] Pidgeon, KC; Bragg, S; Ball, K; et al. Uncommon cause of death: the use of Taser guns in South Florida. *J Emer Nur*. 2008, 34, 305-307.
[14] Braidwood Commission of Inquiry. *Restoring public confidence: restricting the use of conducted energy weapons in British Columbia.* Victoria, British Columbia: Braidwood Commission on Conducted Energy Weapon Use. 2009.
[15] American Medical Association. *Use of Tasers by law enforcement agencies, report 6 of the Council on Science and Public Health 6-A-09.* Washington, DC: American Medical Association, Reference Committee D. 2009.
[16] Eastman, AL; Metzger, JC; Pepe, PE; et al. Conductive electrical devices: a prospective, population-based study of the medical safety of law enforcement use. *J Trauma: Inj Infect Crit Care.* 2008, 64, 1567-1572.
[17] Smith, MR; Kaminski, RJ; Alpert, GP; et al. *A multi-method evaluation of police use of force outcomes: Final report*. Columbia, SC: Univ. of South Carolina. 2008.
[18] MacDonald, JM; Kaminski, RJ; Smith, MR. The effect of less-lethal weapons on injuries in police use-of-force events. *Amer J Pub Health.* 2009, 99, 1-7.

3. Cardiac Rhythm Issues

There is currently no medical evidence that CEDs pose a significant risk for induced cardiac dysrhythmia in humans when deployed reasonably. The heart rhythm issues most important to consider are ventricular fibrillation (VF), ventricular capture (pacing), ventricular tachycardia (VT), atrial fibrillation and pulseless electrical activity (PEA).

Based on research in swine, the risk of CEDs directly causing ventricular fibrillation is exceedingly low.[1-4] VF is more or less likely depending on the energy vector, i.e., where the darts of the CED are located relative to the heart. Different vectors appear to have lesser or greater chance of producing VF with the greatest risk in swine being sternal notch to heart apex or sternal notch to just above the umbilicus (navel).[4]

There is one case report in the medical literature documenting VF two minutes after the collapse of a teenager who was subdued with a CED.[5] The proximity of this collapse to CED use and documented VF argues in favor of an electrically induced cardiac event. A recent review of in-custody deaths associated with CED use evaluated individuals who collapsed within 15 minutes of exposure.[6] Presenting rhythms were available in 56 subjects. In 52 subjects bradycardia-asystole or PEA was seen. The rhythm was VF in four subjects (7 percent). Only one patient collapsed within one minute of exposure, as would typically be expected with VF. Two had a more delayed collapse at five to eight minutes, and one collapsed before exposure. In-custody deaths rarely occur immediately following use of the device, but occur more typically minutes to hours later.[7] Because a VF-related death would be expected to be almost immediate, VF is unlikely to be the cause in most of these in-custody deaths.

There are telemetry and echocardiographic data in swine to demonstrate rapid ventricular capture (pacing) from CED use with a transcardiac vector (when the darts are located on either side of the heart).[8-11] In some of these animals the ventricular dysrhythmia did not terminate with the end of CED discharge and at times led to the death of the animal. The risk of ventricular capture also appears to be dependent on the vector.[12] There are echocardiographic studies in humans during CED activation, one of which has dart placement in the chest area over the heart that did not show capture.[13-15] All other echo studies in humans had remote dart placement and did not show capture.[16] In human studies, the CED exposure is typically applied using alligator clips. Subcutaneous dart placement — such as often occurs during a law enforcement use-of-force incident — is rarely used. Because device output through alligator clips is typically lower in energy, human studies may not reflect the full range of cardiac CED exposures. There are recent studies of rhythm analysis just before, during and after CED discharge showing no sustained dysrhythmia.[17-19]

Rapid ventricular pacing is a method used by electrophysiologists to induce ventricular tachycardia, and this may potentially lead to ventricular fibrillation minutes later. While VT may be pulseless, patients can sometimes be hemodynamically stable for a period of minutes to hours. In other words, a CED may induce rapid ventricular pacing or VT in an individual who appears to be in satisfactory condition, but this may lead to VF after a short delay. Currently, there are no documented cases that CEDs have caused this sequence of events in humans, but it is theoretically possible.

The risks of cardiac arrhythmias or death remain low and make CEDs more favorable than other weapons. Extended CED discharge(s) in swine where rapid ventricular pacing occurred has (have) led to death in some of these animals.[20] Therefore, it cannot be concluded that extended discharge in humans is always safe, despite the successful outcomes of extended discharges documented in the literature.

Pacemakers are implantable cardiac devices that maintain heart rhythm when it gets too slow. Defibrillators are implantable cardiac devices that can function as pacemakers, but are designed to detect life-threatening rapid rhythms and shock or stop the abnormal rhythm. There have been anecdotal, though well-documented, examples of cardiac capture by CEDs in subjects with implantable cardiac devices. In no case, however, were these events associated with bad outcomes.[21-23] There is a case report of an individual with an implanted pacemaker demonstrating ventricular capture during CED use.[21] It cannot be known if the presence of the pacemaker or its associated wires facilitated capture in the ventricle. In swine studies, capture has occurred in the absence of internal wires. An ultrasonographic study did not replicate this finding in human volunteers,[24] and data from field experience does not indicate that complications from capture by CEDs are common.[25-27]

Nonetheless, CED use on individuals with pacemakers and defibrillators can be potentially hazardous. Pacing may be inhibited or asynchronous during CED exposure.[28-29] There has not been a documented case in which a pacemaker has undergone a power-on reset or triggered an elective replacement indicator (which may be associated with pacemaker malfunction). Additionally, there has not been a documented case in which CED exposure caused a long-term change in pacemaker function, such as lead sensing or pacing threshold. Implantable cardiac defibrillators have been demonstrated to detect CED discharges as potential ventricular fibrillation and have charged but not activated.[23,28] Limiting the duration of CED discharges will minimize the chance that one of these devices will give an inappropriate shock.

Risk of ventricular dysrhythmias is exceedingly low in the drive-stun mode of CEDs because the density of the current in the tissue is much lower in this mode. However, there is a case report in the literature where a patient documented to be in atrial fibrillation became combative and was subdued with one drive stun delivered directly over the heart. He was immediately documented to be in a sinus rhythm thereafter.[30] An individual's heart rhythm can spontaneously convert from atrial fibrillation to sinus (normal) rhythm. Nonetheless, the conversion from atrial fibrillation to a sinus rhythm in this case would appear to be temporally attributable to the CED.

In approximately one-quarter of CED deployments in the field the darts strike the anterior chest.[31] With dart deployment the most likely vector to produce cardiac effect would be near the heart and in line with the long axis of the heart.[12,31] Deployments to other regions of the body are very unlikely to generate enough current in the region of the heart to cause ventricular capture or fibrillation. Additionally, when subjects are exposed to CED deployment in the field they often fall and may land in a prone position, driving darts further into the chest wall. This decrease in dart-to-heart distance may increase the likelihood of direct cardiac effects. Individuals of smaller stature may have a shallower distance between the skin and the heart, so they may be more susceptible to cardiac effects associated with dart placement near the heart. This possibility is of theoretical concern and has not been demonstrated.

There is a multitude of ECG and cardiac enzyme data in the literature supporting no significant long-term effects on the heart by CED use. Autopsies have not demonstrated evidence of myocardial infarction (heart attack). The available data do not show long-term blood chemistry changes affecting cardiac function. There are some recent data demonstrating significant increase in blood acidity (acidosis) in animal models after CED use.[21] Some research has examined the role of exertion in combination with CED effects. Extreme physical exertion causes an increase in acidosis because of the production of lactate in the muscles. Severe acidosis can cause spontaneous dysrhythmias that would not be a direct effect of CED use.[32] Additionally, severe acidosis can lead to pulseless electrical activity which may be a mechanism of sudden death seen after a prolonged struggle. CED exposure does not appear to worsen the acidosis that is present from exertion alone.[33-35] Metabolic effects of CED exposure are detailed elsewhere in this report.

There is a controversial case report of the successful resuscitation of a teenager with bipolar disorder and polysubstance abuse who was subdued with a CED. He was reportedly found not to be moving approximately 20 minutes after CED exposure. Emergency medical services personnel found him to be in asystole shortly thereafter. The individual was resuscitated and eventually discharged from the hospital with no apparent long-term deficits.[36] In one publication, bradycardia-asystole or PEA was seen in 93 percent of sudden deaths which quickly followed discharge of CEDs[6]. Either of these dysrhythmias can be precipitated by severe acidosis or could be the terminal rhythm following another life-threatening rhythm. It remains unclear if CED use contributes to the development of PEA or asystole. Rapid recognition of a possibly reversible dysrhythmia in cases like this is imperative to allow for attempted resuscitation.

Although sudden death occurs in custody with and without the use of CED, the exact mechanism of death in many cases is often not clear.[7,37,38] Sometimes, individuals who have been restrained or are in the process of being subdued will stop moving or responding. In many cases, the individual may simply be passively compliant. In some cases, the individual may be experiencing a medical emergency related to acidosis, respiratory compromise, or cardiac arrythmia. Therefore, the restrained individual should be constantly monitored for responsiveness and general medical condition.

Conclusions and Recommendations

Law enforcement personnel are trained to target center body mass when using CEDs. TASER® International, Inc., (a major CED manufacturer) has recently recommended a change in target zone to below the chest. TASER® Bulletin 15 states, "By simply lowering the preferred target zone by a few inches to lower center mass, the goal of achieving Neuro Muscular Incapacitation (NMI) can be achieved more effectively while also improving risk management."[39] The panel does recognize that CED use involving the area of the chest in front of the heart area is not totally risk-free; current research does not support a substantially increased risk of cardiac dysrhythmia in field situations from anterior chest CED dart penetrations.

References

[1] Webster, JG; Will, JA; Sun, H; et al. *Can Tasers® directly cause ventricular fibrillation?* Madison, WI: University of Wisconsin. 2007.

[2] Wu, JY; Nimunkar, AJ; Sun, H; et al. Ventricular fibrillation time constant for swine. *Physio Meas*. 2008, 29, 1209-1219.

[3] Wu, JY; Sun, H; O'Rourke, A; et al. Taser blunt probe dart-to-heart distance causing ventricular fibrillation in pigs. *IEEE Trans Biom Eng*. 2008, 55, 2768-2771.

[4] Lakkireddy, D; Wallick, D; Verma, A; et al. Cardiac effects of electrical stun guns: does position of barbs contact make a difference? *Pacing Clin Electrophysiology*. 2008, 31, 398-408.

[5] Kim, PJ; Franklin, WH. Ventricular fibrillation after stun-gun discharge. *New Eng J Med*. 2005, 353, 958-959.

[6] Swerdlow, CD; Fishbein, MC; Chaman, L; et al. Presenting rhythm in sudden deaths temporally proximate to discharge of TASER conducted electrical weapons. *Acad Emer Med.* 2009, 16, 726-739.

[7] Ho, JD; Heegaard, WG; Dawes, DM; et al. Unexpected arrest-related deaths in America: 12 months of open source surveillance. *West JEM.* 2009, 10, 68-73.

[8] Nanthakumar, K; Billingsley, IM; Masse, S; et al. Cardiac electrophysiological consequences of neuromuscular incapacitating device discharges. *J Amer Coll Cardiol.* 2006, 48, 798-804.

[9] Nanthakumar, K; Masse, S; Umapathy, K; et al. Cardiac stimulation with high voltage discharge from stun guns. *Can Med Assoc J.* 2008, 178, 1451-1457.

[10] Walter, RJ; Dennis, AJ; Valentino, DJ; et al. TASER X26 discharges in swine produce potentially fatal ventricular arrhythmias. *Acad Emer Med.* 2008, 65, 1478-1487.

[11] Valentino, D; Walter, R; Dennis, A; et al. TASER discharges capture cardiac rhythm in a swine model. *Acad Emer Med.* 2007, S104.

[12] Walter, RJ; Dennis, AJ; Valentino, DJ; et al. Taser X26 discharges in swine: ventricular rhythm capture is dependent on discharge vector. *Acad Emer Med.* 2008, 15, 66-68.

[13] Ho, JD; Reardon, R; Lapine, A; et al. *Echocardiographic determination of cardiac rhythm during trans-thoracic wireless conducted electrical weapon exposure.* Minneapolis, MN: Hennepin County Medical Center. n.d.

[14] Ho, JD; Reardon, RF; Dawes, DM; et al. *Ultrasound measurement of cardiac activity during conducted electrical weapon application in exercising adults.* Sorrento, Italy: The Fourth Mediterranean Emergency Medicine Congress. 2007.

[15] Ho, JD; Reardon, RF; Dawes, DM; et al. Echocardiographic evaluation of a TASER-X26 application in the ideal human cardiac axis. *Acad Emer Med.* 2008, 15, 838-844.

[16] Ho, JD; Dawes, DM; Reardon, R; et al. *Cardiac & diaphragm ECHO evaluation during TASER device drive stun.* Minneapolis, MN: Hennepin County Medical Center. 2008.

[17] Vilke, GM; Sloane, C; Levine, S; et al. Twelve-lead electrocardiogram monitoring of subjects before and after voluntary exposure to the Taser X26. *Amer J Emer Med.* 2008, 26, 1-4.

[18] Vilke, G; Sloane, C; Bouton, K; et al. Cardiovascular and metabolic effects of the Taser on human subjects. *Acad Emer Med*. 2007, 14, S104-S105.

[19] Bozeman, WP; Barnes, DG; Winslow, JE; et. al. Immediate cardiovascular effects of the TASER X26 conducted electrical weapon. *Emer Med J*. 2009, 26, 567-570.

[20] Dennis, AJ; Valentino, DJ; Walter, RJ; et al. Acute effects of TASER X26 discharges in a swine model. *J Trauma Inj Infect Crit Care*. 2007, 63, 581-590.

[21] Cao, M; Sinbane, JS; Gillberg, JM; et al. Taser-induced rapid ventricular myocardial capture demonstrated by pacemaker intracardiac electrograms. *J Cardiol Electrophysiology*. 2007, 18, 876-879.

[22] Marine, J. Stun guns: a new source of electromagnetic interference for implanted cardiac devices. *Heart Rhythm*. 2006 3, 342-344.

[23] Haegeli, LM; Sterns, LD; Adam, DC; et al. Effect of a Taser shot to the chest of a patient with an implantable defibrillator. *Heart Rhythm*. 2006, 3, 339-341.

[24] Ho, JD; Dawes, DM; Reardon, RF; et al. Echocardiographic evaluation of a TASER X26 application in the ideal human cardiac axis. *Acad Emer Med*. 2008, 15, 838-844.

[25] Bozeman, WP; Teacher, E. Incidence and outcomes of transcardiac TASER probe deployments. *Acad Emer Med*. 2009, 16, S196.

[26] Bozeman, WP. Additional information on TASER safety. *Ann Emer Med*. 2009, 54, 758-759.

[27] Swerdlow, CD; Fishbein, MC; Chaman, L; et al. Presenting rhythm in sudden deaths temporarily proximate to discharge of TASER conducted electrical weapons. *Acad Emer Med*. 2009, 16, 726-39.

[28] Lakkireddy, D; Khasnis, A; Antenacci, J; et al. Do electrical stun guns (TASER-X26?) affect the functional integrity of implantable pacemakers and defibrillators? *Eur Soc Cardiol*. 2007, 9, 551-556.

[29] Khaja, A; Govindaraja, G; McDaniel, W; et al. Effect of stun gun discharges on pacemaker function. *Circ*. 2008, 118, S592.

[30] Richards, KA; Kleuser, LP; Kluger, J. Fortuitous therapeutic effect of a Taser shock for a patient in atrial fibrillation. *Ann Emer Med*. 2008, 52, 686-688.

[31] Bozeman, WP; Hauda II, WE; Heck, JJ; et al. Safety and injury profile of conducted electrical weapons used by law enforcement officers against criminal suspects. *Ann Emer Med*. 2008, 20, 1-10.

[32] American Heart Association. 2005 American Heart Association guidelines for cardiopulmonary resuscitation and emergency cardiovascular care: Part 7.2: Management of cardiac arrest. *Circ.* 2005, 112, 58-66.

[33] Ho, JD; Dawes, DM; Cole, JB; et al. Lactate and pH evaluation in exhausted humans with prolonged TASER X26 exposure or continued exertion. *Forensic Sci Int.* 2009, 190, 80-86.

[34] Vilke, GM; Sloan, CM; Suffecool, A; et al. Physiologic effects of the TASER after exercise. *Acad Emer Med.* 2009, 16, 704-10.

[35] Ho, JD; Dawes, DM; Buttman, LL; et al. Prolonged TASER use on exhausted humans does not worsen markers of acidosis. *Amer J Emer Med.* 2009, 27, 413-418.

[36] Schwarz, ES; Barra, M; Liao, MM. Case report : successful resuscitation of a patient in asystole after a TASER injury using a hypothermia protocol. *Amer J Emer Med.* 2009, 27, 515, e1-e2.

[37] Cevik, C; Otabachi, M; Miller, E; et al. Acute stress cardiomyopathy and deaths associated with electronic weapons. *Int J Cardiol.* 2009, 132, 312-317.

[38] Samuels, MA. The brain-heart connection. *Circ.* 2007, 116, 77-84.

[39] *Memo regarding Training Bulletin 15.0 regarding medical research update and revised warnings.* Scottsdale, AZ: TASER International, Inc. 2009.

4. Respiratory and Metabolic Issues

The balance of acid and base in the body is maintained by the respiratory system and the kidneys. These respond to the metabolic demands of the individual. As with rigorous exercise, the CED causes muscle contractions that produce lactate in the blood. Lactate lowers the pH of blood, making it more acidic. Respiratory rates increase to counteract this effect by reducing the amount of carbon dioxide (CO_2) in the blood and thereby mitigating the effects of the increased lactate. In extreme cases, the increase in blood acidity (referred to as "acidosis") could lead to cardiac arrest. Studies of CED effects have examined respiration, blood chemistry and the effects on muscle groups. In particular, observation of persons subjected to CED exposure seems to indicate that muscle groups are affected that fall outside those in the area directly between the darts. For example, CED discharges to the thorax often result in collapse to the ground, suggesting that there may be a spinal cord reflex involved that can affect muscle groups under the control of lower spinal cord levels. If that is the case, it seems reasonable that intercostal (between the ribs) muscles used for respiration could also be impacted, with an adverse effect on ability to breathe during CED exposure.

Research to date, however, shows that human subjects seem to maintain the ability to breathe during exposure to a CED. In fact most evidence suggests hyperventilation with an increase in respiratory rate, tidal volume, and minute ventilation during CED exposure. Direct observation of diaphragmatic movement was seen in one study.[1] Despite the hyperventilation, which typically produces an increase in blood pH, a mild decrease in pH indicating metabolic acidosis is often seen with more prolonged exposures. In conjunction with this is an increase in lactate consistent with metabolic acidosis. Alcohol consumption appears to contribute only minimally to an additional decrease in pH or increase in lactate levels.[2]

Very little research has been done on the role of CED vectors (i.e., the positioning of the CED darts) and the effect on respiration. Some studies have examined variable vectors, but with a focus on cardiac effects. As noted below, it is difficult to examine respiratory effects in animal studies.

A recent study of 104 volunteers reports that 18 percent of subjects with CED exposure to the back perceived an inability to breathe during CED exposure, but such inability to breathe was not documented by direct observation or physiologic tests of breathing capacity. The researchers concluded that the results pointed mainly to a need for further study. The medical panel reviewed an unpublished follow-up study using sensors to monitor breathing directly. That study appears to indicate that CEDs could interfere with the ability to inhale, depending on dart placement. Breathing is controlled by the phrenic nerve, which originates in the cervical spinal cord and innervates the diaphragm, in conjunction with intercostal nerves, which originate in the thoracic spinal cord and innervate the intercostal muscles. Therefore, if CED exposure interferes with breathing, it may not be an all-or-none phenomenon. For example, the intercostal muscles may be affected while the diaphragm is not, or vice versa. Further study with objective measurement of breathing is needed to draw more definitive conclusions. Such studies should involve both short term CED exposures and more prolonged or repeated exposures. Hypoventilation could contribute a respiratory component to any underlying acidosis. With prolonged exposure, if CO_2 levels rose significantly, respirations could be further suppressed from the high CO_2 levels despite termination of CED exposure.

Studies with swine have been conducted using an extended exposure of 80 seconds, producing significant acidemia as well as hypoventilation. A few of these animals have died. The animal literature is complicated by the use of sedation that may play a role in hypoventilation and a failure of respiratory compensation for a metabolic acidosis. In other words, the animals' breathing may be compromised by some combination of sedation, CED exposure and other confounding factors from the experimental design. Animal studies suggest that the metabolic acidosis is secondary to an increase in lactate produced after strenuous muscle contraction. In one study, animals were paralyzed to prevent muscle contraction during CED exposure. In this case, acidosis was much less severe but significant cardiac effects were still observed.[3]

There are recent data in the literature of human studies looking at the effect of exercise and CED exposure and their individual contributions to blood acidosis. CED exposure does not appear to add to acidosis above and beyond that seen with exercise to exhaustion. CED exposure without exertion produces only a mild acidosis.[4-6]

Conclusions and Recommendations

Significant acidosis can lead to pulseless electrical activity and may be a mechanism of sudden death in custody. Of particular concern is the possible role that systemic acidosis may play in addition to any metabolic abnormalities or drug intoxication seen in excited delirium, as discussed elsewhere in this report. Further study is required in this area. Until the role of CEDs with respect to respiration has been researched fully, it would be appropriate for law enforcement personnel, when possible, to refrain from continuous activations of longer than 15 seconds. In any case, it is recommended that the medical condition of the individual be constantly monitored during and after CED exposure, regardless of the duration of exposure.

In addition to the concerns related to the effect of CED exposure on respiration, there is a case report in the literature of pharyngeal (throat) perforation from CED discharge.[7] This patient presented with spitting of blood and difficulty breathing.

REFERENCES

[1] Ho, JD; Dawes, DM; Reardon, R; et al. *Cardiac & diaphragm ECHO evaluation during TASER device drive stun*. Minneapolis, MN: Hennepin County Medical Center. 2008.
[2] Moscati, R; Ho, J; Dawes, D; et al. Physiologic effects of prolonged conducted electrical weapon discharge on intoxicated adults. *Acad Emer Med*. 2007, 14, 63-64.
[3] Walter, RJ; Dennis, AJ; Valentino, DJ; et al. TASER X26 discharges in swine produce potentially fatal ventricular arrhythmias. *Acad Emer Med*. 2008, 15, 66-73.
[4] Ho, JD; Dawes, DM; Cole, JB; et al. Lactate and pH evaluation in exhausted humans with prolonged TASER X26 exposure or continued exertion. *Forensic Sci Int*. 2009, 190, 80-86.
[5] Vilke, GM; Sloane, CM; Suffecool, A; et al. Physiologic effects of the TASER after exercise. *Acad Emer Med*. 2009, 16, 704-710.
[6] Ho, JD; Dawes, DM; Buttman, LL; et al. Prolonged TASER use on exhausted humans does not worsen markers of acidosis. *Amer J Emer Med*. 2009, 27, 413-418.
[7] Al-Jarabah, M; Coulston, J; Hewin, D. Pharyngeal perforation secondary to electrical shock from a Taser gun. *Emer Med J*. 2008, 25, 378.

5. CEDs as Contributors to Stress

"Stress," as used in this discussion, describes the body's reaction to threat or physical insult, including but not limited to the adrenaline-related (adrenergic or catecholamine) "flight or fight" reaction. The literature on the acute and chronic effects of stress is large and will not be reviewed extensively here.

Whenever law enforcement officers subdue or restrain an individual, they are contributing to the person's stress level. All aspects of an altercation (including verbal altercation, flight, physical struggle, or physical restraint) constitute stress that may heighten the risk of sudden death, generally from a cardiac dysrhythmia. Whether or not a CED deployment is involved and regardless of the intent of the officer, it is possible for the actions of an officer to directly or indirectly contribute to death by inducing stress. Stress induced by the criminal action of others may be considered a contributing factor in initiating the mechanism of death in certain individuals with underlying natural disease. For example, if an individual with a heart condition dies as a result of being the victim of a robbery, the death may be ruled a homicide caused by the stress of the crime[1-3]. In a similar fashion, stress may be an important issue to consider when investigating and certifying deaths following CED use or when other forms of restraint or subdual are used. One proposed mechanism by which CED use may contribute to death is by increasing stress, which can potentiate the adrenergic responses of tachycardia (i.e., rapid heart rate) and elevated blood pressure, making it an issue related to cause and manner of death determination. There may also be additional physiologic or metabolic effects, especially when stress is severe or other factors have already put the individual into a compromised medical condition, as may occur in individuals who have pre-existing cardiac or other significant disease or who are intoxicated. An important question is whether or not stress caused by CED exposure is different enough from other forms of stress during the agitation, restraint or subdual to justify its separate consideration when certifying death.

The data used to address the stress issue have been derived largely from prospective studies conducted on human volunteers. Medical research suggests that a single exposure of less than 15 seconds deployed from a TASER® model X26™ or a similar model CED is not a stress of a magnitude which separates it from the other stress-inducing components of restraint or subdual.[4] There were no cardiac dysrhythmias among healthy volunteers exposed to one discharge of a TASER® model X26™ for less than 15 seconds following either anaerobic exercise, rigorous exercise or exercise to exhaustion.[4-6] A study using drive-stun mode on volunteers also failed to show cardiac rhythm disturbances or diaphragm disturbances.[7] However, because the numbers of subjects in these studies were small, the subjects were healthy, and the risk of ventricular fibrillation due to a single CED discharge is very low, the applicability of these studies to field conditions is questionable.

It has been proposed that acute stress can damage the heart muscle. There are several reports that suggest that acute stress (with catecholamine release) may cause a cardiomyopathy (or disease of the heart muscle) and be induced in certain individuals during police confrontation. There are insufficient data to provide diagnostic criteria for such a syndrome, although some research and case reports exist.[8-11] Japanese cardiologists initially described “acute stress cardiomyopathy” with transient left ventricular apical ballooning and normal coronary vessels in otherwise healthy, asymptomatic individuals who died in police custody.[8] Such deaths occurred in the absence of CED exposure and are believed to involve a sudden cardiac dysrhythma induced by a surge in adrenaline. Other studies of CED exposure have examined parameters such as blood chemistry, cardiac enzymes and blood gases.[5,12,13] Although studies on human volunteers undergoing prolonged (greater than 15 second) CED exposure showed statistically significant changes in blood gases, these changes (or any respiratory impairment) appear to have limited clinical significance in these healthy individuals.[13,16]

Further study is needed to determine the quantity of stress caused by prolonged or repetitive CED exposure in normal subjects, and larger numbers of human subjects need to be tested. Similar studies in persons with significant disease or drug intoxication would provide more useful data. However, it is not ethical to conduct human studies which attempt to replicate certain "field conditions" (such as drug intoxication with agitation) encountered in CED-associated, police confrontation deaths. The fatal mechanisms of stress and catecholamine release need further clarification, and methods to measure and quantify stress effects should be investigated. Until such methods are developed or more comprehensive field data are obtained, it is reasonable to infer that the effects of acute stress can be cumulative, and that the cumulative effects of adrenaline and other factors such as acidosis may increase an individual's risk of experiencing a sudden cardiac dysrhythmia.

Conclusions and Recommendations

Current data on stress induced by CED exposure are limited because the number of persons studied (sample size) is small and the subjects typically have been healthy volunteers. Further, interpretations are hampered because reliable markers for catecholamine-related stress and its complications are not well identified or accepted. Cases of death may exist where the CED deployment may be the only or predominant inducer of stress. Special attention to such cases is warranted when considering potential mechanisms of death.

CED exposure may contribute to "stress," and stress may be an issue related to cause-of-death determination. All aspects of an altercation (including verbal altercation, physical struggle or physical restraint) constitute stress that may heighten the risk of sudden death in individuals who are intoxicated or who have pre-existing cardiac or other significant disease. Medical research suggests that CED deployment during restraint or subdual is not a contributor to stress of a magnitude that separates it from the other stress-inducing components of restraint or subdual.[15]

References

[1] Hanzlick, R; Hunsaker, JC; Davis, GJ. *A guide for manner of death classification.* Atlanta, GA: National Association of Medical Examiners. 2002.

[2] Davis, JH. Can sudden cardiac death be murder? *J Forensic Sci.* 1978 23, 384-387

[3] Turner, SA; Barnard, JJ; Spotswood, SD; Prahlow, JA. "Homicide by heart attack" revisited. *J Forensic Sic.* 2004, 49, 598-600

[4] Dawes, D; Ho, J; Miner, J. *The neuroendocrine effects of the TASER X26: A brief report.* Minneapolis, MN: Hennepin County Medical Center. 2009.

[5] Vilke, GM; Sloane, CM; Neuman, T; et al. In reply ... physiological effects of the Taser. *Ann Emer Med.* 2008, 52, 85.

[6] Ho, J; Dawes, D; Calkins, H; et al. Absence of electrocardiographic change following prolonged application of a conducted electrical weapon in physically exhausted adults. *Acad Emer Med.* 2007, 14, 128-129.

[7] Ho, JD; Dawes, DM; Reardon, R; et al. *Cardiac & diaphragm ECHO evaluation during TASER device drive stun.* Minneapolis, MN: Hennepin County Medical Center. 2008.

[8] Cevik, C; Otabachi, M; Miller, E; et al. Acute stress cardiomyopathy and deaths associated with electronic weapons. *Int J Cardiol.* 2009, 132, 312-317.

[9] Samuels, MA, The brain-heart connection, *Circ.* 2007, 116, 77-84.

[10] Wittstein, IS; Thiemann, DR; Lima, JA; et al. Neurohumoral features of myocardial stunning due to sudden emotional stress. *N Eng J Med.* 2005, 352, 539-548.

[11] Martínez-Sellés, M. Sudden death in young males after police detention: A new syndrome of possible cardiovascular origin. *Rev Esp Cardiol.* 2009, 62, 101-102.

[12] Valentino DJ, Walter RJ, Dennis AJ, et al. Acute effects of MK63 stun device discharges in miniature swine. *Mil Med.* 2008, 63, 581-590.

[13] Ho, JD; Dawes, DM; Miner, JR. *Serum biomarker effect of prolonged TASER XREP device exposure.* Minneapolis, MN: Hennepin County Medical Center. 2008.

[14] Nixon, AA. Police take Taser training; Littleton officers get to use the devices. *The Caledonian-Record News*. 2007, 1-4.
[15] Bouton, K; Vilke, G; Chan, T; Sloane, C; Levine, S; Neuman, T; Levy, S; Kolkhorst, F; FACSM. *Physiological Effects of a Five Second TASER Exposure. Medicine & Science in Sports & Exercise,* 2007, 39, S323.
[16] Dawes, DM; Ho, JD; Johnson, MA; Lundin, E; Janchar, TA; Miner, JR. 15-second conducted electrical weapon exposure does not cause core temperature elevation in non-environmentally stressed resting adults. *Forensic Sci Int*. 2008, 176(2-3), 253-257.

6. Excited Delirium

Excited delirium (ExD) is one of several terms that describe a syndrome that is broadly characterized by agitation, excitability, paranoia, aggression, great strength and unresponsiveness to pain, and that may be caused by several underlying conditions, frequently associated with combativeness and elevated body temperature.[1-3]ExD-associated agitated behavior often leads to law enforcement intervention and CED use. The predominant theory of the underlying etiology of ExD is an excess of catecholamines (such as adrenaline) or sympathetic nerve stimulation during the excited period. However, a syndrome, by definition, is a collection of signs and symptoms, not a specific disease. People with multiple conditions may present in this manner, including drug-induced psychosis, serotonin syndrome, diabetic ketoacidosis, paranoid schizophrenia and others. Alcohol withdrawal and head trauma have also been implicated.[4] Recent research suggests that individuals with a history of chronic illicit stimulant abuse may be particularly susceptible to excited delirium.[5]

There has been criticism of the term “excited delirium” because its use is generally limited to medical examiners and emergency medicine physicians whose patients die before a complete workup is completed that would allow for a more specific diagnosis. Whether one uses the term or not, ExD-related behavior and medical conditions are well-recognized.

In general, excited delirium may have a mortality of about 10 percent.[6] Sympathomimetic agents include substances such as cocaine, methamphetamine, epinephrine (adrenalin), and dopamine. There is a subset of ExD-affected people who have sympathomimetic poisoning with malignant hyperthermia (high body temperature), sometimes associated with elevated serotonin levels. These cases have a grim prognosis and are at high risk of death regardless of police actions or method of subdual. In one study of 12 patients who made it to the hospital, four died and five suffered severe neurologic complications. This correlates well with other published observations that mortality is about 67 percent for those with a temperature above 41.5 degrees Celsius (106.7 degrees Fahrenheit).[6,7] ExD is frequently but not always associated with the use of cocaine and other stimulants.[8] One study reported that 78 percent of excited delirium cases had serological evidence of stimulant intoxication.[9]

There are other forms of combative, agitated behavior that require subdual; often grouped together under the umbrella of emotionally disturbed persons (EDPs). EDPs may be mistaken for people with excited delirium, and a subset of these may in fact display features of ExD. However, not all EDPs that require subdual have the syndrome of ExD.

There is ongoing research in how best to manage patients with ExD. However, it is clear that at least some of these patients are medically unstable and in a rapidly declining state with a risk of mortality in the short term. This holds true even with medical intervention or in the absence of CED deployment or other types of subdual. While studies in young, healthy, drug-free volunteers suggest that CED deployment has inconsequential metabolic and stress-related effects, no human studies have been performed in situations modeling ExD.[10]

Because of this uncertainty, the number and duration of the CED discharge(s) should be generally limited to the minimal amount needed to attain restraint. Police officers should be aware of ExD-related behavior and indications, especially hyperthermia, which is easy to recognize and associated with the worst outcomes.

Conclusions and Recommendations

The "drive-stun" or contact mode of CED use is a pain compliance procedure, and does not cause muscular incapacitation enabling restraint. Some sources indicate that people suffering from excited delirium are relatively insensitive to pain as a result of their condition. Some reports from law enforcement reinforce this view, because there are individuals who do not appear to be affected by the pain associated with CED exposure. Thus, "drive-stun" mode and other pain compliance methods should not be repeated in these individuals if they are found to have little or no initial effect.

REFERENCES

[1] American College of Emergency Physicians. Excited Delirium Task Force White Paper Report to the Council and Board of Directors. 2009.

[2] Dimaio, TG; Dimaio, JM. *Excited delirium syndrome: Cause of death and prevention*. CRC Press, 2006.

[3] *Report of the Panel of mental health and medical experts' review of excited delirium*. Halifax, Nova Scotia, Canada Department of Justice, 2009.

[4] Pacquette, M. Excited delirium: Does it exist? *Persp Psychiatric Care*. 2003, 39, 93-94.

[5] Mash, DC; Duque, L; Pablo, J; et al. Brain biomarkers for identifying excited delirium as a cause of sudden death. *Forensic Sci Int*. 2009, 190, e13-e19.

[6] Samuel, E; Williams, RB; Ferrell, RB. Excited delirium: Consideration of selected medical and psychiatric issues. *Neuropsychiatric Dis Treat*. 2009, 5, 61-66.

[7] Gowing, LR; et al. The health effects of ecstasy: A literature review. *Drug Alcohol Rev*. 2002, 21, 53-63.

[8] Stratton, SJ; Rogers, C; Brickett, K; et al. Factors associated with sudden death of individuals requiring restraint for excited delirium. *Amer J Emer Med*. 2001, 19, 187-191.

[9] Robison, D; Hunt, S. Sudden in-custody death syndrome. *Topics Emer Med*. 2005, 27, 36-43.

[10] Strote, J; Hutson, HR. Taser use in restraint-related deaths. *Prehospital Emer Care*. 2006 ,10, 447-450.

7. Safety Margins of CEDs

Most fatalities involving CED use are in people who have other risk factors for sudden death. This is a concern for law enforcement, because a large number of arrestees will have unrecognized clinical states of drug intoxication or pre-existing medical conditions that put them at risk for sudden, unexpected death, regardless of the type of subdual or restraint used. The medicolegal death investigator must identify the currently recognized safety margins of CED deployment in order to evaluate competing possible causes of death. Most of the deaths reviewed by the panel for this report involved individuals with drug intoxications or complicating medical conditions or both, thus making judgments about the relative role of CED exposure in the deaths very difficult.

It is clear that physical injury secondary to dart puncture, fall and other physical effects is a real though relatively uncommon danger. These are discussed at length elsewhere in the report, as is the literature regarding the cardiac, respiratory and metabolic effects of CED use. The latter suggest small risks associated with CED use, especially for healthy individuals.

However, there are groups who may be at risk for sudden death and those who are more vulnerable to physical insult. These disparate but occasionally overlapping groups include small children, those with diseased hearts, the elderly and pregnant women. For instance, the death of a seven-month-old infant following the application of a stun gun by his foster mother has been reported.[1] The small size of this infant, coupled with the nearness of the contact electrodes to the heart, was postulated as a plausible mechanism for death. Case reports of fetal death due to exposure to electrical current exist, all involving exposure significantly more severe than that associated with CED exposure.[2] In contrast, one study of 31 pregnant women subjected to electric shock, not from CED deployment, but including 12 V (telephone line), 110 to 220 V (home appliance), and 2000 and 8000 V (electric fence) current, found no adverse effects to the pregnancies.[3] There has been no research or field study demonstrating a significantly higher or lower risk for CED use with any particular group.[4-7]

Unlike the risk of secondary injury due to falling or puncture, the risk of human death due directly or primarily to the electrical effects of CED application has not been conclusively demonstrated. However, there are anecdotal cases where no other significant risk factor for death is known and where the temporal association provides circumstantial evidence of causation, albeit weak.[8] The panel recognizes the distinction between correlation and causation and that close temporal relationships do not necessarily prove causation. Studies on ventricular fibrillation with respect to dart placement, demonstration of ventricular fibrillation, pulseless ventricular tachycardia, or pulseless electrical activity in animals, and anecdotal examples of ventricular capture in humans with cardiac pacemakers or defibrillators provide a plausible mechanism for unusual and rare cases of death due to a confluence of unlikely circumstances. Multiple plausible mechanisms have been proposed but none proven.[9]

Many subjects of CED exposure are under the influence of drugs. One study suggested that cocaine intoxication decreased the risk of arrhythmia in animals, though it was limited by the lack of controls and the complex manipulation of the animals required by the study.[10] Similarly, a study on prolonged exposure in alcohol-intoxicated adult humans revealed no significant morbidity.[11] Thus, there is currently no basis in scientific research to conclude that drug use increases or decreases the safety margin of CED exposure.[11]

The safety margin of CEDs is subject to the variability in the output of the devices. Researchers are continuing to study the most common CEDs in use today, the models X26™ and M26™ from TASER® International, Inc., to determine the variability of their output. The effect of this output variability on cardiac safety margin is unclear.

Most research has been done using devices from TASER® International, Inc. Medical and safety data regarding stun batons, CED projectiles and other devices are much more limited. Although the early data suggest similar results, the current literature is sparse.[12-16] Another manufacturer, Stinger Systems, Inc., manufactures CEDs that are being used in some agencies and that are purported to have an improved safety margin because they declare to operate at lower power levels than the TASER® models X26™ or M26™. Independent research on Stinger Systems devices is very limited, so the panel could not judge the relative safety margin of these devices.[17]

Conclusions and Recommendations

The literature suggests a substantial safety margin with respect to the use of CEDs when they are used according to manufacturer's instructions. However, plausible mechanisms of injury do exist which make it impossible to exclude direct lethality in every case. The safety margins of CED use in normal healthy adults may not be applicable in small children, those with diseased hearts, the elderly, pregnant women and other potentially at-risk individuals. The effects of CED exposure in these populations are not clearly understood, and more data are needed. The use of a CED on these individuals when recognized during attempted subdual should be minimized or avoided unless the situation excludes other reasonable options.

The use of manual techniques, baton blows, CEDs, other less-lethal technologies and even taking no action at all will each carry its own risks. All evidence suggests that the use of CEDs carries with it a risk as low as or lower than most alternatives. While it should be remembered that unlikely events may occur, it is unreasonable to demand that any application of force be totally risk-free in all populations at all times. The decision to use a CED or other options is best left to the reasonable tactical judgment of trained law enforcement at the scene.

REFERENCES

[1] Turner, MS; Jumbelic, ML. Stun gun injuries in the abuse and death of a seven-month old infant. *J Forensic Sci.* 2003, 48, 180-182.

[2] Jaffe, R; Fejgin, M; Aderet, B. Fetal death in early pregnancy due to electric current. *Acta Obstet Gynecol Scand.* 1986, 65, 283.

[3] Einarson, A; Bailey, B; Inocencion, G; et al. Accidental electric shock in pregnancy: a prospective cohort study. *Am J Obstet Gynecol.* 1997, 176, 678-681.

[4] Wu, JY; Sun, H; O'Rourke, AP; et al. *Dart-to-heart distance when TASER® causes ventricular fibrillation in pigs.* International Federation for Medical and Biological Engineering Proceedings. 2007, 15, 1-5.

[5] Wu, JY; Sun, H; O'Rourke, A; et al. Taser blunt probe dart-to-heart distance causing ventricular fibrillation in pigs. *IEEE Trans Biomed Engineering*, 2008, 55, 2768-2771.

[6] Webster, JG; Will, JA; Sun, H; et al. *Can Tasers® directly cause ventricular fibrillation?* Madison, WI: University of Wisconsin. 2007.

[7] Whitehead, S. Sorting Taser truths from Taser mythology. *Lauren County Emergency Medical Services: Our Newsletter.* 2006, 1, 13-14.

[8] Swerdlow, CD; Fishbein, MC; Chaman, L; et al. Presenting rhythm in sudden deaths temporally proximate to discharge of TASER conducted electrical weapons. *Acad Emer Med.* 2009, 16, 726-739.

[9] Cevik, C; Otabachi, M; Miller, E; et al. Acute stress cardiomyopathy and deaths associated with electronic weapons. *Int J Cardiol.* 2009, 132, 312-317.

[10] Lakkireddy, D; Wallick, D; Ryschon, K; et al. Effects of cocaine intoxication on the threshold for stun gun induction of ventricular fibrillation. *J Am Coll Cardiol.* 2006, 48, 805-811.

[11] Moscati, R; Ho, J; Dawes, D; et al. Physiologic effects of prolonged conducted electrical weapon discharge on intoxicated adults. *Acad Emerg Med.* 2007, 14, 63-64.

[12] Valentino, DJ; Walter, RJ; Dennis, AJ; et al. Acute effects of MK63 stun device discharges in miniature swine. *Mil Med.* 2008, 173, 167-173.

[13] Vilke, G; Sloane, C; Bouton, K; et al. Cardiovascular and metabolic effects of the Taser on human subjects. *Acad Emer Med.* 2007, 14, S104-S105.

[14] Dawes, DM; Ho, JD; Lundin, E; et al. *The effects of the eXtended range electronic projectile (XREP) on breathing.* Minneapolis, MN: Hennepin County Medical Center, 2008.

[15] Dawes, DM; Ho, JD; Johnson, MA; et al. *Breathing parameters, venous blood gases, and serum chemistries with exposure to a new wireless projectile conducted electrical weapon in human volunteers.* Minneapolis, MN: Hennepin County Medical Center. 2007.

[16] Burdett-Smith, P. Stun gun injury. *J Accident Emer Med.* 1997, 14, 402-404.

[17] Mesloh, C; Henych, M; Thompson, LF; et al. *A qualitative & quantitative analysis of conducted energy devices: TASER X26 vs. Stinger S200.* Washington, DC: U.S. Department of Justice, National Institute of Justice. 2008.

8. Prolonged Exposure

There is no evidence in animals that indicates a high risk of injury from a single discharge lasting less than 15 seconds from a TASER® X26™. Unlike the TASER® X26™, which requires the user to hold the trigger to maintain discharges longer than five seconds, other CEDs will apply a longer discharge without any intervention from the user. The TASER® C2™, designed for civilian use, applies a 30-second exposure to a target. Thirty-second exposure to the output of the TASER® C2™ CED in swine resulted in significant changes in blood chemistry, although most of the blood changes returned to baseline after the CED discharge ended. This raises concern for potential detrimental effects due to use of the TASER C2™ CED.[1] However, in one study, 20-to 30-second C2™ CED application in healthy humans had no significant deleterious effects on their physiology.[2]

The most common version of the dart-mode CED is the X26™ manufactured and sold by TASER® for law enforcement. When the trigger is pulled and the darts attach to the skin or clothing, the device delivers its standard charge as an initial pulse wave of up to 50 kV, followed by a series of low-current (2.1 milliamps, 70 mJ) pulses for five seconds. The device has the ability, however, to deliver extensively prolonged and uninterrupted discharges. The standard discharge cycle may be shortened or prolonged by either maintaining pressure on the trigger *continuously* over variable periods of time or by *repeatedly* depressing and releasing the trigger over variable intervals limited only by the power in the battery (approximately five minutes).

There is no standard definition of "prolonged" CED exposure for either continuous duration or number of multiple interrupted discharges. The majority (93 percent) of CED exposures in the field involve 15 seconds or less; a significant body of the medical literature has employed 15 seconds or less of CED exposure.[3]

After a review of anecdotes that seemed to indicate that multiple exposures were more hazardous, one researcher recommended in 2005 — without supporting documentation — that law enforcement agents should "... [l]imit the number of TASER® exposures when possible (3 is probably a reasonable number)."[4] The Police Executive Research Forum produced guidelines for police concerning CED use including a recommendation that "[w]hen activating a CED, law enforcement officers should use it for one standard cycle and stop to evaluate the situation (a standard cycle is five seconds). If subsequent cycles are necessary, agency policy should restrict the number and duration of those cycles to the minimum activations necessary to place the subject in custody."[5] The Canadian Police Research Centre recommended: "... continuous cycling of the TASER for periods exceeding 15-20 seconds may increase the risk ... and should be avoided where practical."[6]

Recommendations by the principal manufacturer, TASER® International Inc., have changed over time. Prior to 2008, they warned against extended duration applications [greater than 5 seconds], noting in particular that darts over the chest or diaphragm may impair respiration and cautioned that "... [u]sers should avoid prolonged, extended, uninterrupted discharges or extensive multiple discharges whenever practicable...."[7] Their 2008 training bulletin (#14) concludes that more recent tests on humans demonstrate that "... there are no adverse effects on heart function or respiration deriving from multiple or prolonged deployments.[8]

Studies examining the effects of extended exposure in humans to CEDs are limited to humans exposed to less than 45 seconds. The majority of studies are limited to exposures of 15 seconds or less. Review of deaths following CED exposure indicates that some are associated with prolonged or multiple discharges of the CED. By contrast, experiments using healthy human volunteers have found no cardiac dysrhythmias[9,10] or respiratory dysfunction[11] following exposures less than 45 seconds. There are no published studies of humans exposed in excess of 45 seconds. Continuous 15 second application of the X26™ to either the back or chest of "physically exhausted" adult humans (designed to mimic field situations), over a 12-inch anatomic spread encompassing the heart, yielded normal electrocardiograms.[13]

Bozeman et al. reported in 2008 that among 1,201 cases in which a CED was used, 18.5 percent received CED discharges three or more times.[13] In one of these 222 incidents, an individual sustained significant injury, although it is unclear whether the CED played a role in the injury. The repeated or continuous exposure of a CED to an actively resisting individual may not achieve compliance, especially when the individual may be under drug intoxication or in a state of excited delirium.

The medical risks of repeated or continuous CED exposure beyond the durations studied in humans are currently unknown, and the role of CEDs in causing death is unclear in these cases. Uncertain risks associated with the effect of CEDs on respiration should be noted, as detailed elsewhere in this report (see chapter 4). These risks reinforce the view that prolonged, continuous CED exposure should be avoided, if possible.

Despite the well recognized limitations implicit in the applicability of results of animal experiments to humans, the evidence from experiments with swine models indicates that repeated exposures of over 80 to 90 seconds total duration have been associated with increased risk of ventricular fibrillation and mortality.[14-16] Swine studies involving exposure durations of 15 seconds or less are not associated with increased risks for ventricular fibrillation.[17] Intermittent exposures appear to be tolerated better than continuous exposure.[15-19]

Conclusions and Recommendations

There may be circumstances in the field that require repeated or continuous exposure to a CED discharge. Law enforcement personnel should be aware that the associated risks are unknown and that most deaths associated with CED use involved multiple or prolonged discharges. Therefore, multiple or prolonged activations of CED as a means to accomplish subdual should be minimized or avoided.

REFERENCES

[1] Jauchem, JR; Seaman, RL; Klages, CM. Physiological effects of Taser C2 conducted energy weapon. *Forensic Sci Med Path.* 2009, 5, 189-198.

[2] Ho, JD; Dawes, DM; Cole, CB; et al. *Human physiological effects of a civilian conducted electrical weapon application.* Minneapolis, MN: Hennepin County Medical Center. n.d.

[3] Bozeman, WP; Hauda II, WE; Heck, JJ; et al. Safety and injury profile of conducted electrical weapons used by law enforcement officers against criminal suspects. *Ann Emer Med.* 2009, 53, 480-489.

[4] Czarnecki, F. *Taser use recommendations for law enforcement officers.* Miami Beach, FL: International Association of Chiefs of Police. 2005.

[5] Police Executive Research Forum. *Conducted energy device policy and training guidelines for consideration.* Washington, DC: PERF Center on Force and Accountability. 2005.

[6] Manojlovic, D; Hall, C; Laur, D; et al. *Review of conducted energy devices, Technical Report TR-012006.* Ottawa, Canada: Canadian Police Research Service. 2005.

[7] TASER International, Inc. *Restraint during TASER™ system application. Training Bulletin 12.004, TASER Law Enforcement Warnings.* Scottsdale, AZ: TASER International, Inc. 2005.

[8] TASER International, Inc. *Restraint during TASER™ system application. Training Bulletin 14.003, TASER Law Enforcement Warnings.* Scottsdale, AZ : TASER International, Inc. 2008. From Braidwood Commission. *Restoring public confidence: restricting the use of conducted energy weapons in British Columbia.* Vancouver, British Columbia, Canada. The Braidwood Commission on Conducted Energy Weapon Use. 2009.

[9] Vilke, GM; Sloane, CM; Bouton, KD; et al. Physiological effects of a conducted electrical weapon on human subjects. *Ann Emer Med.* 2007, 50, 569-575.

[10] Moscati, R; Ho, J; Dawes, D; et al. Physiologic effects of prolonged conducted electrical weapon discharge on intoxicated adults. *Acad Emer Med.* 2007, 14, 63-64.

[11] Dawes, DM; Ho, JD; Lundin, E; et al. *The effects of the eXtended range electronic projectile (XREP) on breathing.* Minneapolis, MN: Hennepin County Medical Center. 2008.

[12] Ho, J; Dawes, D; Calkins, H; et al. Absence of electrocardiographic change following prolonged application of a conducted electrical weapon in physically exhausted adults. *Acad Emer Med.* 2007, 14, 128-129.

[13] Bozeman, WP; Hauda II, WE; Heck, JJ; et al. Safety and injury profile of conducted electrical weapons use by law enforcement officers against criminal suspects. *Ann Emer Med.* 2008, 20, 1-10.

[14] Dennis, AJ; Valentino, DJ; Walter, RJ; et al. Acute effects of TASER X26 discharges in a swine model. *J Trauma Inj Infect Crit Care*. 2007, 63, 581-590.

[15] Walter, RJ; Dennis, AJ; Valentino, DJ; et al. TASER X26 discharges in swine produce potentially fatal ventricular arrhythmias. *Acad Emer Med*. 2008, 15, 66-73.

[16] Valentino, D; Walter, R; Dennis, A; et al. TASER discharges capture cardiac rhythm in a swine model. *Acad Emer Med*. 2007, S104.

[17] Jauchem, JR; Cook, MC; Beason, CW. Blood factors of Sus scrofa following a series of three TASER electronic control device exposures. *Forensic Sci Int*. 2008, 175, 166–70.

[18] Jauchem, JR; Sherry, CJ; Fines, DA; et al. Acidosis, lactate, electrolytes, muscle enzymes, and other factors in the blood of Sus Scrofa following repeated TASER exposures. *Forensic Sci Int*. 2006, 161, 20-30.

9. Research Associated with the Decision to Use a CED

Law enforcement agencies have deployed CEDs under a variety of circumstances and with a range of agency policies. The determination of appropriate use-of-force in police action has an extensive literature that goes well beyond the scope of this panel. There are currently efforts at a national level to establish guidelines for use within this context.[1-3] Individual departments revise their policies on a continuing basis. In one study of more than 500 agencies, 14.9 percent of agencies surveyed indicated that they were considering changing their use-of-force policies, and 21 percent already had.[4] Some agency policies allow the use of a CED only as an alternative to deadly force. In many cases, policies permit the use of CEDs in a wider variety of incidents, including passive resistance scenarios.[5] Among other considerations, agencies must consider the safety aspects of CED deployment when making these policy decisions. In addition, medical examiners are commonly called upon to offer an opinion about the level of force that was applied in a custody-related death. The recognition of appropriate versus inappropriate use of force can have significant medicolegal consequences.

It was not the mandate of this panel to develop use-of-force policies for law enforcement agencies or to review CED-related deaths with respect to whether police acted appropriately in any specific instance or whether specific policies or force options are advisable. Nonetheless, it is clear that the relative risk associated with CED deployment must be viewed in relationship to the risks of other alternatives, and not viewed in a vacuum. Multiple departmental reviews have suggested that injury rates, death rates and complaints against police drop significantly following the deployment of CEDs. For instance, deployment of CEDs in Charlotte, N.C., was associated with a 56.4 percent reduction in officer injury and a 79 percent reduction in suspect injury.[6] An independent study has indicated an increase in in-custody deaths following the adoption of CEDs, based on survey data, but the role of CEDs in any of these deaths was not examined.[7] These results are not normalized for crime rates or other factors.

Independent studies of use-of-force outcomes involving CEDs have been completed, and they substantiate the view that CED deployment, in general, decreases the likelihood of injuries to suspects and officers.[5,8-10] Further, national statistical data indicates that, despite widespread use of CEDs in law enforcement, CED deployment is associated with only a small proportion of in-custody deaths.[11] In the largest independent study to date, involving 12 agencies and more than 24,000 use-of-force cases, the odds of suspect injury decreased by almost 60 percent when a CED was used.[8,9] Officer injuries were either unaffected or reduced when a CED was used. In contrast, using physical force increased the odds of injury to officers by more than 300 percent and to suspects by more than 50 percent.[8,9] In general, the outcome data are consistent with medical research and this panel's review of deaths following CED deployment. Deployment of CED has a margin of safety as great as or greater than most alternatives.[12-14]

Conclusions and Recommendations

In general, CEDs are safe when used properly. Nonetheless, care should be taken when CEDs are deployed. Researchers have recommended that passive resisters should not be subjected to CED use and that CED discharges should be limited to the number needed to gain control of the suspect.[8-10] It has been suggested that CEDs should not be used unless the only other alternative is lethal force. However, if a goal is minimization of harm, it is appropriate to use the force application that is associated with the least likelihood of injury. CED use is associated with a significantly lower risk of injury than physical force, so it should be considered as an alternative in situations that would otherwise result in the application of physical force. Police officers need to be aware that, although CEDs provide an effective alternative to lethal force, it is still possible to misuse the device if it is deployed outside the bounds of departmental policies derived from national guidelines. Use-of-force policies are a function of training, cultural context, operational contingencies and scientific concerns. Beyond the recognition of the lower injury rates to officers and suspects associated with CED use, it was not the mandate of this panel to make recommendations for a national use-of-force model or precisely where CED use should be placed within it.

REFERENCES

[1] Cronin, JM; Ederheimer, JA. *Conducted energy devices: Development of standards for consistency and guidance*. Washington, DC: U.S. Department of Justice, Office of Community Oriented Policing Services, 2006.

[2] International Association of Chiefs of Police. *Electronic control weapons. Model policy # 64.* Alexandria, VA: International Association of Chiefs of Police. 2008.

[3] International Association of Chiefs of Police. *Electro-muscular disruption technology: A nine-step strategy for effective deployment.* Alexandria, VA: International Association of Chiefs of Police. 2008.

[4] Alpert, GP; Dunham, R. *Understanding police use of force: Officers, suspects, and reciprocity.* Cambridge, England: Cambridge University Press. 2004.

[5] Smith, MR; Kaminski, RJ; Rojek, J; et al. The impact of conducted energy devices and other types of force and resistance on officer and suspect injuries. *Policing: Int J Police Strat Manage*. 2007, 30, 423-446.

[6] Charlotte-Mecklenburg Police Department. *TASER project first-year full deployment study.* Charlotte, NC: Police Department. 2005.

[7] Lee, BK; Vittinghoff, E; Whiteman, D; et al. Relation of Taser (electrical stun gun) deployment to increase in in-custody sudden deaths. *Amer J Cardiol*. 2009, 103, 877-880.

[8] MacDonald, JM; Kaminski, RJ; Smith, MR. The effect of less lethal weapons on injuries in police use-of-force events. *Amer J Public Health.* 2009, 99, 1-7.

[9] Smith, MR; Kaminski, RJ; Alpert, GP; et al. *A multi-method evaluation of police use of force outcomes.* Columbia, SC: University of South Carolina. 2009.

[10] Taylor, B; Woods, D; Kubu, B; et al. *Comparing safety outcomes in police use-of-force cases for law enforcement agencies that have deployed conducted energy devices and a matched comparison group*

that have not: a quasi-experimental evaluation. Washington, DC: Police Executive Research Forum. 2009.
[11] Mumola, CJ. *Arrest-related deaths in the United States, 2003-2005.* Washington, DC: U.S. Department of Justice, Bureau of Justice Statistics. 2007.
[12] Eastman, AL; Metzger, JC; Pepe, PE; et al. Conductive electrical devices: A prospective, population-based study of the medical safety of law enforcement use. *J Trauma: Inj Infect Crit Care*. 2008, 64, 1567-1572.
[13] Bozeman, WP; Hauda II, WE; Heck, JJ; et al. Safety and injury profile of conducted electrical weapons use by law enforcement officers against criminal suspects. *Ann Emer Med.* 2008, 20, 1-10.
[14] Jenkinson, E; Neeson, C; Bleetman, A. The relative risk of police use-of-force options: evaluating the potential for deployment of electronic weaponry. *J Clin Forensic Med.* 2006, 13, 229-241.

10. Post-Event Medical Care

Individuals who have received CED discharges may suffer injuries during the incident and also may have pre-existing medical conditions or traumatic injuries, which should be assessed by medical personnel. Appropriate medical care should be provided if these are present or suspected, especially when falls, burns or other trauma occur, or when darts penetrate obviously sensitive areas of the body.

Medical Screening. Some form of medical screening is recommended after all CED exposures starting at the scene of the incident. This may take the form of jail intake medical screening, evaluation by emergency medical service (EMS) providers in the field, or by hospital emergency department personnel.

Dart Removal. In most cases, darts embedded in the skin may be removed at the scene by properly trained medical or law enforcement personnel in accordance with local protocols. When removing embedded darts, care should be taken to avoid exposure to bloodborne pathogens. Individuals handling darts should be mindful of sharp points and additional spines located around the components of the newer CED device projectiles. Medical care should be provided when darts are located in potentially vulnerable areas such as the face, eyes, neck, genitals or groin, or if there is concern for underlying injuries, regardless of body location.[1-4] Monitoring in-custody. Ongoing monitoring of suspects while in custody is strongly recommended. Changes in physical condition or mental status/behavior may occur due to effects of drugs (which may have been ingested or undergone continued absorption), medical conditions, or as a result of head trauma or internal injuries. These subjects should be immediately referred for medical evaluation and appropriate therapy delivered by qualified specialists.

Outpatient Follow-Up. In the absence of injuries, no specific medical follow-up is required after most CED exposures. However, suspects who have an implanted cardiac device (pacemaker or implanted defibrillator) should be evaluated by a physician and have the device and its stored data analyzed.[5] In cases with ocular injuries or CED discharge near the eyes, outpatient ophthalmologic follow-up is recommended to exclude complications such as retinal detachment or delayed cataract formation.[2,6] Those reporting or suspected of having significant medical or psychiatric conditions following CED use should also be evaluated to determine if they may be CED-related and to provide appropriate care. Although neuropsychologic dysfunction and complaints (physical, cognitive and emotional) have been well-documented with non-CED electrical injury, it is not clear at this time if this may also occur after CED exposure.[7]

Continued Abnormal Behavior. A minority of suspects taken into police custody (with or without CED use) will exhibit continued or ongoing abnormal behavior. Abnormal mental status and/or increased body temperature in combative or resistive subjects may be associated with an increased risk for sudden cardiac arrest and death. Underlying medical or drug-induced conditions (such as hypersympathomimetic states, hyperthermia, acidosis, excited delirium, rhabdomyolysis and others) may be responsible for extensive struggling and other behaviors that require subdual by law enforcement, including the use of CEDs. There could also be underlying changes in body chemistry, hypoxia and/or acidosis due to suspect behavior and activities prior to subdual and CED use.[8] Precautions should be taken during any form of restraint to allow for reasonable chest movement and airway protection.[9]

Abnormal agitation and confusion should be treated by law enforcement personnel as a medical emergency. EMS should be immediately dispatched to the scene when this is recognized (law enforcement should not wait until a subject is subdued and in custody; EMS should be called immediately). Further, it must be recognized that a nonmoving or unresponsive subject may be in a medical crisis (i.e., cardiac arrest) rather than being intentionally passive.

Emergency medical treatment. In such cases, emergency medical providers should initiate medical support as soon as it is safe to do so. If warranted, sedation, hydration and cooling should be provided as soon as possible in addition to standard assessment, resuscitation and supportive care. Emergency medical services protocols specifying these interventions in the field may be useful and are already in place in some systems.[10]

Medical personnel both in the field and in the hospital setting are encouraged to assess and document vital signs including body temperature and oxygen saturation levels, cardiac rhythm,[9,11] neurologic status, and physical findings. Spinal precautions and diagnostic evaluations for traumatic injuries may be appropriate based on the history and physical findings. Blood and urine samples should be obtained early for laboratory studies, which may include serum glucose, electrolytes, pH, lactate levels, cardiac enzymes, urine toxicology screen and urine myoglobin, among others.[12,13]

Forensic aspects of medical care, Some agencies obtain photographs of imbedded CED darts in the field prior to removal. In cases of critical illness, injuries or death, all darts and clothing removed during medical care (after photography prior to removal if feasible) should be retained for investigative purposes by the medical examiner/coroner/law enforcement agency and handled as evidence. Detailed records of medical treatment should be maintained in all cases.

Conclusions and Recommendations

Medical personnel should provide appropriate care to individuals who have received CED discharges as these individuals may suffer injuries during the incident and may also have preexisting medical conditions needing assessment. Medical screening at the scene of the incident, the proper removal of dart(s), and the ongoing monitoring of individuals in custody for abnormal physical and behavior changes are crucial procedures. Suspects with implanted cardiac devices should receive outpatient follow-up as necessary. Detailed records, including photographs of the scene and body, should be obtained in all cases; these records should include documentation of medical treatment provided.

REFERENCES

[1] Han, JS; Chopra, A; Carr, D. Ophthalmic injuries from a TASER. *J Canadian Assoc Emer Physicians*. 2009, 11, 90-93.

[2] Chen, SL; Richard, CK; Murthy, RC; et al. Perforating ocular injury by Taser. *Clin Exper Ophthal*. 2006, 34, 378-380.

[3] Al-Jarabah, M; Coulston, J, Hewin, D. Pharyngeal perforation secondary to electrical shock from a Taser gun. *Emer Med J*. 2008, 25, 378.

[4] Rivera-Garcia, LE; Crown, LA; Smith, RB. Overview of electronic weapon injury and emergency department management. *Amer J Clin Med*. 2008, 5, 46-49.

[5] Haegeli, LM; Sterns, LD; Adam, DC; et al. Effect of a Taser shot to the chest of a patient with an implantable defibrillator. *Heart Rhythm*. 2006, 3, 339-341.

[6] Seth, RK; Abedi, G; Daccache, AJ; et al. Cataract secondary to electrical shock from a Taser gun. *J Cataract Refract Surg*. 2007, 33, 1664-1665.

[7] Pliskin, NH; Capelli-Schellpfeffer, M; Law, RT; et al. Neuropsychological symptom presentation after electrical injury. *J Trauma: Inj Infect Crit Care*. 1998, 44, 709-715.

[8] Robison, D; Hunt, S. Sudden in-custody death syndrome. *Topics Emer Med*. 2005, 27, 36-43.

[9] Strote, J; Hutson, HR. Taser use in restraint-related deaths, *Prehospital Emer Care*. 2006, 10, 447-450.

[10] ACEP Excited Delirium Task Force. *White paper report on excited delirium syndrome. Proceedings of the American College of Emergency Physicians Council Meeting,* Irving, TX: American College of Emergency Physicians. 2009.

[11] Stratton, SJ; Rogers, C; Brickett, K; et al. Factors associated with sudden death of individuals requiring restraint for excited delirium. *Amer J Emer Med*. 2001, 19, 187-191.

[12] Tsai, SH; Chu, SJ; Hsu, CW; Cheng, SM; Yang, SP. Use and interpretation of cardiac troponins in the ED. *Amer J Emer Med*. 2008, 26, 331-341

[13] Pidgeon, KC; Bragg, S; Ball, K; et al. Uncommon cause of death: the use of Taser guns in South Florida. *J Emer Nur*. 2008, 34, 305-307.

11. Considerations in Death Investigation

If a death occurs following the use of a CED by law enforcement personnel who are subduing, restraining, or apprehending a subject, the death will be investigated by the appropriate medical examiner or coroner's office as an in-custody death. Because deaths following CED deployment involve both complex and predictable issues, the death investigation needs to include consideration of information that may not be gathered in a routine death investigation or other in-custody death investigations. It is not the intent of this report to provide a comprehensive checklist of tasks which should be performed. Rather, we are providing what we believe will be helpful suggestions for consideration in the most important aspects of CED-related death investigations.

The information needed for investigation of death following CED use will need to be collected by death investigators from multiple sources and at the direction of the medical examiner or coroner who has ultimate responsibility for determining the cause and manner of death in the case. Further, the forensic pathologist who performs the autopsy will need to review such information, perhaps request additional information, and will develop information from the autopsy examination which may trigger or require additional investigation. The forensic pathologist who performs the autopsy is an integral part of the investigative team.

The following information can be useful in establishing facts and should be considered during the death investigation:

1. A timeline of all events with attempts to verify, to the extent possible, the accuracy of the dates and times of reported events, with specific emphasis on the interval between CED use, unresponsiveness and death.
2. Clarification of CED model and mode of use (drive-stun and/or cartridge mode).
3. Access to a comparable CED for familiarization with design and functionality;
4. Recent activities of the subject prior to the incident.
5. The emotional state of the subject.
6. The subject's reaction to each deployment.
7. The subject's medical conditions as determined by medical history, medical record review and medical conditions determined at autopsy.

8. The subject's drug use history, including prescription and illicit drugs as well as alcohol.
9. Specific inquiry into the subject's cardiac history, including review of any electrocardiograms or other cardiac function or laboratory tests which have been performed in the past.
10. Specific inquiry into the subject's seizure history to rule out history of seizures or to clarify the nature of a past seizure disorder.
11. Review of witness accounts, police reports, use-of-force reports, emergency medical services records, medical and psychiatric records, and any videos, photographs or digital images of the events.
12. Determination whether body temperature and ambient temperature were established and documentation of dates and times of such recordings.
13. If death occurred after arrival at a hospital, obtaining blood drawn upon arrival at the hospital so it may be tested for intoxicants, including medications, if needed.
14. Review of downloaded information from the CED with special attention to an assessment of the number, duration and timing of CED discharges, including correlation with other case information to determine successful delivery and the effects of the discharges on the subject.
15. Assessment of the CED to establish whether it is operating within the manufacturer's specifications.
16. Preservation of the CED with batteries (since removal of batteries may alter the time clock) along with the darts and attached wires.
17. Investigation of the subject's place of residence or last place to visit to determine if additional medical history or evidence of drug use exists.

Assuming that the investigation and autopsy are performed and documented/reported in accordance with the National Institute of Justice's *Death Investigation; A Guide for the Scene Investigator* and the National Association of Medical Examiners' *Forensic Autopsy Performance Standards,*[1,2] additional information and procedures that may be helpful, but not warranted in every case, are as follows:

1. Performance of a complete autopsy of the scope usually performed for deaths in-custody with appropriate histologic sampling of organs.
2. Comprehensive forensic toxicology of autopsy specimens and any retained antemortem samples, specifically including tests for alcohol,

nervous system stimulants, common drugs of abuse, anti-seizure drugs, and therapeutic drugs often prescribed for psychiatric disorders.

3. Measurement of the thickness of the anterior chest wall from the skin to the rear of the pre-pericardial sternum at intercostal space between the left fourth and fifth ribs.
4. Measurement of the thickness of clothing and chest wall or tissue in the area(s) where CED darts or prongs penetrated.
5. Measurement of the depth of dart penetration.
6. Documentation of the CED dart's(s') length(s).
7. Documentation of dart and stun dart locations and any associated marks or burns.
8. Consideration of unusual or atypical current flow paths, such as body to ground, body to water, body to metal, etc.
9. Determination of the nature of any other forms of subdual or restraint that were employed in the case in question.
10. Removal and evaluation (interrogation) of any implanted cardiac or other electronic devices.
11. Utilization of appropriate consultants such as cardiologists, cardiac pathologists and neuropathologists as needed.

The agency responsible for conducting the death investigation should ultimately be responsible for certifying the cause and manner of death.

References

[1] National Medicolegal Review Panel. *Death investigation: A guide for the scene investigator.* Washington, DC: U.S. Department of Justice, National Institute of Justice. 1999.

[2] Peterson, GF; Clark, SC. *NAME forensic autopsy performance standards.* Atlanta, GA: National Association of Medical Examiners. 2006.

12. Considerations in Death Certification

The medical examiner/coroner is required to determine the cause and manner of death in all violent, sudden, and unexpected or unusual deaths.

Consultant experts in various specialties may be involved as the case warrants. Any death related to CED deployment would fit into this category. Available publications describe basic principles regarding death certification and completion of the cause-of-death section of the death certificate (see also the definitions in the Glossary of this report).[1,2] The manner of death classification (homicide, suicide, accident, natural or undetermined) is dependent on autopsy findings in conjunction with all relevant information, including the circumstances surrounding death as determined by a medically objective investigation independent of law enforcement.[3]

In a CED-related death, the medical examiner/coroner may choose to exclude any mention of the CED from the death certificate. In some cases, the death certificate may list the CED as a causative factor in Part I or as a contributory factor (other significant condition) in Part II of the cause-of-death statement. In other cases, the CED may be listed as one of the items in the space provided on the death certificate to describe how injury occurred. Further, the medical examiner/coroner may choose to classify a CED-related death as a homicide, whether the CED itself is directly causative or contributory, because the actions of law enforcement led to the death. In the majority of these cases, a subsequent (nonmedical) investigation would classify the homicide as justifiable, but it is beyond the scope of the medical examiner/coroner to make that determination for a death certificate. In other cases, including those that might list the CED on the death certificate in some way, the death may be ruled an accident, because the judgment of the medical examiner or coroner would be that the actions of law enforcement or others involved did not cause death.

Regardless of these classifications, an independent observer should use caution when interpreting the inclusion of a CED on a death certificate or the classification of the manner of death as a homicide as an absolute indictment of the CED as the sole or primary reason for the death. First, the CED-related deaths examined in this study involved a complex set of circumstances with individuals who were not necessarily healthy and who were often highly drug-intoxicated. These circumstances make it very difficult to point to the CED as a particular cause in specific deaths. Second, the decision to list the CED on the death certificate is subject to the judgment of the individual medical examiner/coroner and includes medicolegal considerations, experience, and often aspects of local practice and history.

Among the medical examiners on the panel that produced this report, many cases resulted in divergent views concerning cause and manner of death, although these disagreements were within the normal bounds of practice among certifiers of death. It is one objective of this report to minimize these differences among medical examiners and coroners by improving the scientific understanding of CED-related injuries and deaths. This is extremely important to medical examiners and coroners who must complete the death certificate and report the cause, manner and circumstances of death, including how injury occurred. A consensus is needed to make certification of death more consistent between cases and between jurisdictions, while always remaining aware of the need for professional judgment.

For deaths in which the subject is in law enforcement custody or is being apprehended, restrained or subdued, the medical examiner/coroner must often determine if the circumstances and findings are most consistent with a natural, accidental, homicidal or undetermined manner of death.

A major problem with the investigation of in-custody deaths and those in which a CED has been deployed is obtaining relevant and accurate information regarding the chronology of events leading up to the time when the subject underwent cardiopulmonary arrest during or following subdual or restraint. A limiting factor is that like all death investigations, in-custody death investigations occur after the fact over extended periods of time following the initial investigation of the scene and circumstances, and often rely on investigative information gathered by the same law enforcement agency involved in the subdual, restraint or deployment of a CED.

Both theoretical and real cases reviewed by the medical panel in which CED deployment was considered as a major factor in causing death were classified as homicide when there were accurate timelines, independent and objective witness accounts, and strong — almost immediate — temporal relationships between CED deployment and death. CED use in these instances could be responsible for initiating or contributing to a fatal sequence of events. It needs to be emphasized that the manner of death classification on a death certificate is not an assessment of legal responsibility for the death. From the medical examiner/coroner standpoint, homicide means that death either occurred at the hands of another person or resulted from hostile, illegal actions or inactions of another person. For example, deaths certified as homicide while in the "care" (i.e., custody) of another person have included the following types of situations:[4]

1. The caregiver has caused the death intentionally.
2. The caregiver lacks required licensure or training for the type of care being provided.

3. The caregiver consciously disregarded a known likelihood of injury and showed a wanton and gross disregard for the well-being of the patient (negligence).

In use-of-force deaths, the actions of law enforcement officers may be judged differently than those of other responders who are classified as "caregivers" even if the officers' actions are very similar to those of emergency medical personnel.

In deaths following CED deployment, a certifier of death may determine that the manner of death was homicide; nonetheless, it may be determined that the officer was acting appropriately and the homicide was justifiable. Alternatively, the prosecuting attorney may pursue homicide charges if the law enforcement officer recklessly engaged in conduct and use of force that created a substantial risk of injury and was not compliant with policy or guidelines of the department (e.g. repetitive CED discharges when the subject has already been restrained and handcuffed, or administration of a CED to a compliant individual). In some cases, an accidental manner of death may be assigned if there is a lethal concentration of drugs or there are lethal complications of drug use, and subdual or CED use are clearly not factors contributing to death. In these cases, when the manner of death is classified as an accident, the certifier of death would be indicating that the actions of the law enforcement officer, whether appropriate or not, did not contribute to the death of the individual.

Certification of death following CED deployment can be difficult because:

- Information needed to draw conclusions may be of poor quality or not available.
- It may be impossible to determine the relative causative or contributory roles of underlying disease, drug intoxication, drug-induced agitation or delirium, restraint or subdual, or possible direct electrical or indirect stresses of CED deployment.

After thorough investigation, the certifier may be reasonably certain that CED deployment did or did not cause or contribute to death. In many cases, the role of CED deployment is much less clear.

There is debate as to whether CED deployment alone can directly cause death in humans via electrical effects on the cardiovascular or nervous system, as has been detailed elsewhere in this report. For the purpose of this discussion it is assumed that such a death may occur. For example, assume a young, thin, healthy person is not intoxicated, but is resisting arrest and receives several intentionally deployed, consecutive CED discharges to the anterior chest, then suddenly dies without other reasonable explanation and no other causative factors are identified. The death certificate could be worded as follows:

Part I	A. **Sudden cardiac death**
	Due to, or as a consequence of: B. **Conducted energy device discharges**
	Due to, or as a consequence of: C.
Part II	OTHER SIGNIFICANT CONDITIONS: Conditions contributing to death, but not resulting in the underlying cause of death in Part I
Manner of Death **Homicide**	Describe how injury occurred **Subdual by law enforcement**

If investigation shows a specific single form of restraint or subdual did cause death, such as head trauma with brain injury from a blow to the head, then death certification may follow this general example:

Part I	A. **Skull fracture with brain contusions**
	Due to, or as a consequence of: **B. Blunt-force head injury**
	Due to, or as a consequence of: C.
Part II	OTHER SIGNIFICANT CONDITIONS: Conditions contributing to death, but not resulting in the underlying cause of death in Part I
Manner of Death **Homicide**	Describe how injury occurred **Struck during subdual by law enforcement for cocaine-induced agitation**

More typically, however, multiple factors are involved such as:

- Repeated or prolonged deployment of the CED.
- Agitated state or delirium.
- Intoxication.
- Use of multiple methods of subdual or restraint.
- Acidosis, hyperthermia or rhabdomyolysis.
- Underlying natural disease such as heart disease, sickle cell trait, etc.

In these less clear-cut cases, the certifier may conclude that subdual contributed to death because of stress, often in conjunction with a drug-induced agitated state or disease. The questions become:

- Should all contributory factors be itemized or should they simply be combined under a general category of "stress of restraint" or "stress of subdual?"
- Would death have occurred when it did without the restraint?
- Should the manner of death be classified as other than homicide?

For example, in a person with cocaine induced agitation and sickle cell trait who the certifier concludes died from subdual, one option for certifying the death is as follows:

Part I	**Cocaine induced delirium resulting in physical subdual**
	Due to, or as a consequence of: B.
	Due to, or as a consequence of: C.
Part II	OTHER SIGNIFICANT CONDITIONS: Conditions contributing to death, but not resulting in the underlying cause of death in Part I Sickle cell trait
Manner of Death Homicide	Describe how injury occurred Cocaine-induced agitation requiring multiple methods of subdual by law enforcement

In many cases, there are multiple forms of subdual or restraint such as carotid sleeper hold, pepper spray, handcuffing, hobbling, "hog-tying'" slaps, asp baton strikes, chest compression, CED deployment, and others. Because it is difficult to differentiate contributory methods from noncontributory ones, and because of limited space in the "how injury occurred" section of the death certificate, it may be best to be generic in these complex cases and simply state that multiple forms of subdual or restraint were used. Of course, if there is reasonable evidence that one or more specific forms of subdual or restraint did cause death, such cases can be certified as described above. In general in these cases, CED deployment should be considered to be a stress of a magnitude that is comparable to other components of subdual.

Many times, law enforcement officers respond to violent or combative subjects and subdue or restrain them in order to facilitate medical care. Often, EMS will request law enforcement officers to come to a scene. In this capacity as a first responder, the distinctions between medical assistance and law enforcement procedures can be blurred. If a fatal injury results during medical assistance, the manner of death is usually classified as an accident. If the fatal injury results during a law enforcement action (even if the motivation is to provide medical assistance), the manner of death may be classified as homicide.

If there is insufficient information to differentiate between two manners of death, the manner of death may be certified as undetermined. Some examples in which an undetermined manner of death may be considered include the following:

a) The autopsy and toxicology findings show no obvious cause of death.
b) Combinations of significant disease and toxicology results that ordinarily would not be fatal.
c) When death is delayed after lengthy hospitalization and circumstantial details are not clear.
d) No toxicology screen was done on admission to the hospital and death is delayed.
e) Circumstances of the incident cannot be accurately determined.

Cases reviewed by the panel where CED was determined to be a major factor, and classified as homicides, were cases in which there was an accurate timeline, an independent witness observation, and strong, almost immediate, temporal relationship between CED use and death or initial/sudden collapse or unresponsiveness. When death or the initial/sudden collapse immediately follows CED use, one can reasonably conclude that the CED would be responsible for initiating a lethal sequence of events.

REFERENCES

[1] U.S Department of Health and Human Services. *Medical examiners' and coroners' handbook on death registration and fetal death reporting.* DHHS Publication No. (PHS) 2003-11110. Hyattsville, MD: Centers for Disease Control and Prevention, National Center for Health Statistics. April 2003.

[2] Hanzlick, R. (ed). *Cause of death and the death certificate: Important information for physicians, coroners, medical examiners, and the public.* Northfield, IL: College of American Pathologists. 2006.

[3] Name: *A guide for manner of death classification.* Marcilene, MO: National Association of Medical Examiners. 2002.

[4] Duncanson, E; Richards, V; Luce, KM; et al. Medical homicide and extreme negligence. *Amer J Forensic Med Path.* 2009, 30, 18-22.

EPILOGUE

The statements, opinions, and recommendations in this report were developed by consensus of the panel members. The opinions of the members may change in the future based on new studies and as more information becomes available. Indeed, the publication of numerous papers in the time between the release of the interim report and this final report was instrumental in determining the final recommendations published here. New data continue to accrue even during the preparation of this final report.

There was a good deal of discussion among the participants regarding the determination of cause and manner of death from a medicolegal viewpoint. Part of the discussion concerned our inability to make dogmatic statements about risk in many of these cases. There were also differing philosophies among participants underlying the placement of specific factors involved in a death within the chain of causation or contribution. As noted in the disclaimer at the beginning of this report, these differences do not reflect basic conceptual differences in the pathophysiology involved, but instead reflect conceptual differences about the meaning of cause and manner of death. In some cases, of course, the determination of cause and manner of death is explicit and noncontroversial. But in cases where the "real" cause must be teased from an interconnecting web of causal factors, differences in opinion will arise. That does not, however, remove the mandate of the medical examiner in most cases to assign a specific cause of death.

In addition to these essentially philosophical issues, the fact is that our knowledge and understanding of CED effects is incomplete. Indeed, there is uncertainty about how exactly CEDs achieve their effects on the human body. Some propose that the effects of CEDs are due entirely to electrically induced tetany, while others hypothesize secondary effects due to nerve stimulation and reflex effects. We do know that CEDs are characterized by the infliction of excruciating pain. While such a thorough comprehension may not be necessary to measure the physiologic effects on cardiac function, metabolism, respiration and mortality associated with CED deployment, it means that all recommendations are subject to revision as our understanding improves.

During discussions of the use of CEDs with stakeholders, interested parties and organizations, a recurring concern arose regarding the use of CEDs as punishment or torture devices. The panel shares the concern that wide deployment of an extremely safe method of delivering extraordinary pain could also potentiate abuse. Questions about the ethical infliction of pain in law enforcement are important, and we applaud efforts to address them, but they are not within the mandate of this panel. Instead, we emphasize that issues of safety are different and should not be conflated with these other important concerns.

The panel extends its deep gratitude to the researchers and stakeholders who shared their knowledge, experience, and extraordinarily diverse perspectives. We greatly appreciate the efforts of the National Institute of Justice in funding and providing logistical support. We thank our respective employers, institutions, universities and our families for allowing us the time and opportunity to perform this function. We extend our respect and thanks to those in law enforcement and the military who protect our lives, liberty and property. We recognize our duty to the citizens of these United States, whom we serve and who deserve our best efforts to ensure that their lives and rights are preserved.

GLOSSARY OF TERMS AS USED IN THIS REPORT

Acidosis — An increase in the acidity (decrease in pH) of the blood; the normal pH of human blood is 7.4.

Adrenergic response — The epinephrine (adrenaline or catecholamine) response to stress such as occurs with the "fight or flight" reaction.

Alligator clip — A small metal clip, which is hinged and has teeth, so it resembles the snout, jaws and teeth of an alligator. In CED research, it is used to attach wires to a research subject's clothing.

Apex (of the heart) — The tip (bottom) of the heart closest to the diaphragm.

Cardiac dysrhythmias (arrhythmias) — Abnormal heart rhythms. These can spontaneously resolve in some instances:

- **Asystole** — Lack of electrical activity and heart function.
- **Atrial fibrillation** — An abnormal heart rhythm where the upper chambers (atria) are fibrillating (quivering in an unsynchronized fashion). The atria fail to augment heart output and often cause the heart to beat very rapidly.

- **Pulseless electrical activity (PEA)** — A state where electrical activity can be recorded from the heart but there is not enough blood flow out of the heart to maintain a pulse or blood pressure.
- **Ventricular capture (pacing)** — The ability of an external source of energy to cause the lower chambers (ventricles) of the heart to beat.
- **Ventricular fibrillation** — An abnormal rapid heart rhythm originating in the lower chambers of the heart. This rhythm does not support flow of blood out of the heart, causing lack of blood pressure or pulse. This rhythm typically leads rapidly to unconsciousness and death.
- **Ventricular tachycardia** — An abnormal rapid heart rhythm originating in the lower chambers of the heart. This rhythm may allow for adequate blood pressure to support life for a period of time, but may also rapidly lead to death.

Cardiac mechanisms — The ways the heart can fail when injured or sick.

Conducted energy device (CED) — A weapon primarily designed to disrupt a subject's central nervous system by means of deploying electrical energy sufficient to cause uncontrolled muscle contractions and override an individual's voluntary motor responses.

Darts — Projectiles that are fired from a CED and penetrate the skin; wires are attached to the darts leading back to the CED.

Dart removal — The act of removing a dart from a person's body or clothing.

Deployment — Making an item available for use in the field or actually using it in the field. In this report, deployment means use of the CED on a subject.

Diabetic ketoacidosis — A metabolic abnormality in diabetics which is characterized by elevated blood sugar and ketones, and may cause abnormal mental function.

Duration — The aggregate period of time that CED shocks are activated.

Dysrhythmia — Any disturbance or irregularity of the heartbeat.

Echocardiography — Ultrasound study of the heart.

Electrocardiogram — A graphic produced by an electrocardiograph, which records the electrical activity of the heart over time.

Electro muscular disruption — The effect that a CED has on the body. Overrides the brain's communication with the body and prevents voluntary control over the muscles.

Emotionally disturbed person (EDP) — A generic term often used by criminal justice and law enforcement personnel to describe a person with behavioral disturbances which may be caused by a mental disorder, disease, or a chemically induced state.

Excited delirium — State of extreme mental and physiological excitement, characterized by extreme agitation, hyperthermia, euphoria, hostility, exceptional strength and endurance without fatigue.

Hypoventilation — Breathing slower or less deeply than normal, thereby increasing the amount of carbon dioxide (CO_2) in the blood to above normal.

Implantable cardiac device — An electronic device surgically implanted in a person and usually consisting of a cardiac pacemaker, defibrillator or combination pacemaker/defibrillator.

- **Implantable cardiac defibrillator** — An implanted cardiac device which has the ability to recognize and treat abnormal rhythms of the heart. This device can function as a pacemaker but is also designed to treat life-threatening rhythms such as ventricular tachycardia and ventricular fibrillation. The device treats these rhythms by either shocking the heart or rapidly pacing the heart back to a normal rhythm.
- **Pacemaker** — An implanted cardiac device which causes the heart to beat when the heart is beating too slow.

Less lethal — A concept of planning and force application that meets an operational or tactical objective,with less potential for causing death or serious injury than conventional, more lethal police tactics.

Less-lethal weapon — Any apprehension or restraint device that, when used as designed and intended, has less potential for causing death or serious injury than conventional police lethal weapons.

Metabolic mechanisms — The ways the metabolism can fail when a person is injured or sick.

Pacing threshold — The amount of energy required from a pacemaker to cause the heart to beat.

Paranoid schizophrenia — A psychotic state in which a person has paranoid delusions (false beliefs or altered perceptions of reality).

Physical nechanisms — The ways in which illness or injury can compromise heart/lung function or put body metabolism at risk.

Pulmonary mechanisms — The ways in which lung function can be compromised by injury or sickness.

Pulse rate — The frequency at which electrical pulse waves are generated.

Pulse wave — A graphic measurement of the wave produced by an impulse of electric energy.

Respiratory — Relating to the act or process of inhaling (breathing in) and exhaling (breathing out); breathing, also called ventilation.

Restrain — To control, limit,or prevent movement.

Restraint — A device that restricts movement.

Rhabdomyolysis — Potentially fatal condition resulting from the breakdown of muscle fibers resulting from metabolic, physical or chemical causes, producing substances that can damage other organs such as the kidneys.

Sensitive areas — A person's head, neck, and genital areas, and a female's breast areas.

Standard CED cycle — A five-second electrical discharge occurring when a CED trigger is pressed and released. The standard five-second cycle may be shortened by turning the CED off. (Note: If a CED trigger is pressed and held beyond five seconds, the CED will continue to deliver an electrical discharge until the trigger is released.)

Sternal notch — The depression in the skin just above the breast bone where the neck connects to the chest.

Subdual — To bring under control.

Sympathomimetic — A chemical agent or physiologic response which mimics or increases bodily responses typically caused by the sympathetic nervous system, often due to agents such as cocaine and amphetamine compounds which increase adrenaline (epinephrine), or neurotransmitters such as dopamine.

Symptomatology — The combined symptoms of a disease: the symptom complex of a disease.

Vector — The angle or course of current in this example.

APPENDIX A. HOW A TASER® CONDUCTED ENERGY WEAPON WORKS

Part 3: Conducted Energy Weapons

Braidwood Commission on Conducted Energy Weapon Use

Models commonly used by Law Enforcement **TASER M26** and **TASER X26**.

a. The Advanced TASER M26

Introduced to the law enforcement community in 1999, the Advanced TASER M26 is a pistol-shaped weapon. It can be used in two modes:

- ***Push-stun mode*** — the end of the weapon is pressed against the target's body (with an expended cartridge attached or without a cartridge attached), and a pulsed electrical current is transferred to the adjacent muscles; or
- ***Probe mode*** — when a cartridge is attached to the end of the weapon, it fires two metal darts or probes (using compressed nitrogen as a propellant), which imbed in the target's skin or clothing. The probes, which have hooked tips, can penetrate up to 9 mm into the subject's skin. If the probes do not reach the skin due to bulky clothing, the high voltage creates an arc enabling the current to enter the body. The probes are connected to the weapon by wires that conduct a pulsed electrical current from the weapon into the target's body.

The trigger activates a five-second electrical current cycle, which can be stopped by placing the safety lever in the safe position, or can be repeated by re-pressing the trigger after the completion of the first cycle. Holding the trigger down continuously can extend a cycle.

Eight AA nickel metal hydride or alkaline cell batteries power the M26. Depending on the battery brand used, the electrical current has a pulse rate of 15 or 20 pulses per second, with a pulse duration of 40 microseconds (40 millionths of a second) full waveform. When the M26 is held level, the upper probe is propelled in a horizontal direction and the lower probe is propelled at an eight-degree downward angle, which means that, for every seven feet of travel, there is a one-foot spread between the probes (or, for every 2.1 metres of travel, there is a 0.3 metre spread). Four different colour-coded single-use cartridges can be installed, with different wire lengths — yellow (15 feet), silver (21 feet), green (25 feet), and orange (35 feet). For the M26 to be effective when used in its probe mode, both probes should hit the subject. To assist the officer in aiming, the M26 emits a red laser beam, which marks where the upper probe will hit the target. Every cartridge has a unique serial number. When it fires out the two probes and wires, it also disperses about 30 small discs, called Anti-Felon Identification tags, with the same serial number on it. This enables investigators to link up the user of the weapon with the person to whom the cartridge was issued. The M26 has an LED indicator showing that the laser is on and the weapon is capable of firing, but it does not indicate whether there is sufficient battery power to fire or discharge. The weapon stores data about firings, date, and time for approximately 585 firings, which can be downloaded using an M26 dataport download kit. The manufacturer's specifications respecting the M26's electrical output, which I will discuss in more detail later, include the following:

- Voltage:
  - Peak open circuit arcing voltage — 50,000 V
  - Peak loaded voltage — 5,000 V
  - Average voltage over duration of main phase — 3,400 V
  - Average voltage over full phase — 320 V
  - Average voltage over one second — 1.3 V
- Current: 3.6 mA average (milliamps)
- Energy per pulse:
  - Nominal at main capacitor — 1.76 joules
  - Delivered into load — 0.50 joules
- Power rating:
  - Nominal at main capacitor — 26 watts at 15 pulses per second
  - Nominal delivered into load — 7.39 watts at 15 pulses per second

However, Mr. Reilly testified that an electrical shock can be delivered across several inches of air and if one probe hits the subject and the other probe falls on wet ground, the subject may still receive a shock.

b. The TASER X26

The manufacturer introduced its X26 model, for law enforcement and military use, in 2003. It was more compact, 60 percent lighter, and designed to be carried in a holster on an officer's service belt. The X26's specifications are similar to the M26, except for the following:

- Batteries — digital power magazine (two 3-volt lithium batteries, as used in digital cameras)
- Pulse rate — 19 pulses per second
- Pulse duration — 100 microseconds (100 millionths of a second)
- Peak loaded voltage — 1,200 V
- Average voltage over duration of main phase — 400 V

- Average voltage over full phase — 350 V
- Average voltage over one second — 0.76 V
- Current — 2.1 mA average
- Energy per pulse:
  - Nominal at main capacitors — 0.36 joules
  - Delivered into load — 0.07 joules
- Power rating:
  - Nominal at main capacitors — 6.84 watts
  - Delivered into load — 1.33 watts
- LED display — a two-digit display of remaining digital power magazine energy percentage, burst time, warranty expiration, unit temperature, illumination status, and current time and date.
- Data storage — stores time, date, burst duration, unit temperature, and remaining digital power magazine energy percentage for approximately 1,500 firings. The data can be downloaded using a USB data interface module.
- Video and audio — available with an optional video and audio recorder that is activated when the safety switch is armed. It is capable of recording for up to 90 minutes.

In order to understand how a conducted energy weapon works, a basic understanding of electricity is required. I am indebted to Mr. J. Patrick Reilly, from the Applied Physics Laboratory of Johns Hopkins University, for his very informative presentation during our public forums. Much of the explanation that follows is based on what he said and his PowerPoint presentation.

To begin with a question, if putting my finger into a 120-volt light socket could kill me, why could I walk away from a 50,000-volt shock from a conducted energy weapon? There are two reasons. First, the "peak open circuit arcing voltage" is rated at 50,000 volts when nothing is connected to the probes, such as when the officer is testing the weapon by creating an electrical arc between the two electrodes. When the weapon is under load (such as when imbedded in a person's skin or clothing), the voltage is much less — 7,000 volts for the M26 and 1,300 volts for the X26, according to Mr. Reilly. Second, the duration of the conducted energy weapon pulse is short. In the case of the wiring in our homes, the electrical current is continuous. However, in a conducted energy weapon, a new electrical pulse begins 19 times every second. The actual duration of each of these pulses is much briefer — 30 microseconds (30 millionths of a second) with the M26 and 80 microseconds

(80 millionths of a second) with the X26. The pulse durations of 30 and 80 microseconds are taken from Mr. Reilly's presentation. According to the manufacturer's specifications, the pulse durations are 40 and 100 microseconds for the M26 and X26 respectively.

There is an important reason why a conducted energy weapon needs 50,000 volts. This voltage (analogous to pressure in a water hose) is required in order to create an electric arc that bridges an air gap. For example, if one of the probes is imbedded in clothing and does not touch the skin, the high voltage creates an arc between the probe and the skin, enabling the electrical current to enter the body. Similarly, although the outer layer of a person's skin (the corneum) is dry and normally a poor conductor, the high voltage breaks down the dryness and makes the skin a good conductor.

Turning now to current (analogous to the water flow rate in a hose, such as litres per minute), the manufacturer's specifications state that the M26 has a current of 3.6 milliamps (3.6 thousandths of an ampere) average, and the X26 has a current of 2.1 milliamps (2.1 thousandths of an ampere) average. Mr. Reilly, on the other hand, cites the M26 as having a peak output current of 17 amperes, and the X26 as having a peak output current of 3 amperes. He explained the difference between his numbers and the manufacturer's numbers as follows. His numbers measure the actual amperage during a pulse, whereas the manufacturer's numbers are an average over the total time period, during and between pulses. In his view, average current is irrelevant to electrostimulation.

According to Mr. Reilly, "delivered charge" is the best indicator of the potential electrostimulation. It is measured in coulombs, which is analogous to the volume of water delivered by a hose during a set period of time. The significant point is that both the M26 and the X26 have an almost identical "delivered charge" for each pulse — approximately 100 micro-coulombs (or 100 millionths of a coulomb). This is so because of the differing currents and pulse durations of the two models, as shown in Table 1.

Table 1. Delivered charge of M26 and X26 models

	M26	*X26*
Current	17 amperes per pulse	3 amperes per pulse
Pulse duration	30 microseconds	80 microseconds

To give a sense of what effect 100 micro-coulombs of delivered charge would have on a person, Mr. Reilly conducted laboratory experiments with human subjects, who were subjected to brief high-voltage pulses on their forearms. Subjects reported pain on average at 0.5 micro-coulombs, and intolerable pain at 1.0 micro-coulombs. This is to be contrasted to the delivered charge of 100 micro-coulombs from each pulse of a conducted energy weapon, which delivers 95 pulses over a five-second period.

The purpose of the electrical current is different, depending on the mode used:

- ***Push-stun mode*** — if the trigger is pulled when the end of the conducted energy weapon is pressed against the person's skin (*e.g.*, arm). The electrodes are close together, which means that the electrical current is localized to the muscles in that area. In that case it serves a pain compliance purpose, to persuade the person to let go of something, or to otherwise comply in order to avoid further shocks.
- ***Probe mode*** — when the probes are deployed they are normally imbedded in the person farther apart than the electrodes are in the push-stun mode. In that case, the electrical current spreads out more and goes deeper into the body, engaging more and more excited tissue. In addition to the same pain experienced in the push-stun mode, the electrical current now interferes with the person's neuromuscular system. The person typically becomes incapacitated, and falls to the ground with no ability to put his or her hands out to break the fall.

When the five-second cycle is over, the pain and/or incapacitation is over, and the person's normal strength returns immediately.

From the Braidwood Commission of Inquiry. *Restoring public confidence: Restricting the use of conducted energy weapons in British Columbia*. Victoria, British Columbia: Braidwood Commission on Conducted Energy Weapon Use. 2009.

APPENDIX B. DEFINITIONS FOR CAUSE, MECHANISM AND MANNER OF DEATH

Background

The study steering group presented definitions for Cause, Mechanism and Manner of Death for review and comment by the Medical Panel in January 2008. The definitions herein were revised in April 2008 and will serve to guide mortality reviews of those cases of interest to the study.

The underlying (or proximate) cause of death is

(a) the disease or injury, or combination of the two, that initiated the pathophysiologic sequence of events leading to death OR
(b) the circumstances of the event [accident or violence] that produced the fatal injury.

The proximate cause of death is always etiologically specific.

The **immediate cause of death** is the terminal disease, injury, medical complication or pathophysiologic condition resulting from the underlying cause or circumstance and directly preceding death.

The underlying cause of death and the immediate cause may either exist simultaneously or be separated by variable spans of time.

An **intermediate (or intervening)** cause of death is a disease or condition with fatal potential that occurs at any time between the underlying cause of death and the immediate cause of death and is a result of the underlying cause.

There may be no, one or multiple intermediate causes of death.

A **contributory cause of death** is any or all significant disease[s], injuries, or pathophysiologic condition[s] that existed at death and that may have fatal potential, but did not lead to or result in the underlying cause of death.

There may be no, one or multiple contributory causes of death.

The **mechanism of death** constitutes the fatal pathophysiologic derangement[s] resulting from the underlying cause of death.

The mechanism of death is one or more complication[s] of the underlying cause of death, and:

- Is a disturbance of physiology and/or biochemistry.

- Is the derangement by means of which the underlying cause of death effects the lethal outcome.
- May have more than one cause.
- Is never etiologically specific.

The **manner of death** is a classification of the circumstances of how death occurred. It is derived from correlation of all investigative and scientific components of the death investigation.

In most jurisdictions in the United States the subdivision of manner of death is as follows:

- **Natural** — Solely due to disease processes.
- **Unnatural (or violent)** — Due to external agencies (injury of any kind, including the toxic effects of chemicals) either exclusively or in concert with natural conditions. These may be:
    - Homicide.
    - Suicide.
    - Accident.
- **Undetermined** — When neither unnatural nor natural manner of death can be determined — OR — if the cause of death is known to be unnatural, but investigation cannot distinguish the subcategories.

Guidelines for Cause (COD) and Manner (MOD) of Death as Used in This Document

Cause and manner of death are the medical opinions of the certifier based on information available at the time of certification.

COD — Reasonable medical and investigative probability, or a preponderance of all scientific and investigative data.

MOD — Reasonable discretion by the investigating certifier, correlating all pertinent case data.

Cause and manner of death are subject to change if new information relevant and material to the investigation emerges.

(N.B. — Certification of a death as homicide does not imply criminal culpability, which is a determination solely in the jurisdiction of the justice system.)

APPENDIX C. THE USE-OF-FORCE CONTINUUM

Most law enforcement agencies have policies that guide their use of force. These policies describe an escalating series of actions an officer may take to resolve a situation. This continuum generally has many levels, and officers are instructed to respond with a level of force appropriate to the situation at hand, acknowledging that the officer may move or skip from one part of the continuum to another in a matter of seconds.

An example of one of many use-of-force continuums follows:

- **Officer Presence — No force is used. Considered the best way to resolve a situation.**
  - The mere presence of a law enforcement officer works to deter crime or diffuse a situation.
  - Officers' attitudes are professional and nonthreatening.
- **Verbalization — Force is not physical.**
  - Officers issue calm, nonthreatening commands, such as "Let me see your identification and registration."
  - Officers may increase their volume and shorten commands in an attempt to gain compliance. Short commands might include "Stop," or "Don't move."
- **Empty-Hand Control — Officers use bodily force to gain control of a situation.**
  - **Soft technique**. Officers use grabs, holds and joint locks to restrain an individual.
  - **Hard technique.** Officers use punches and kicks to restrain an individual.
- **Less-Lethal Methods — Officers use less-lethal technologies to gain control of a situation.**
  - **Blunt impact**. Officers may use a baton or projectile to immobilize a combative person.
  - **Chemical**. Officers may use chemical sprays or projectiles embedded with chemicals to restrain an individual (e.g., pepper spray).
  - **Conducted energy devices (CEDs).** Officers may use CEDs to immobilize an individual. CEDs discharge a high-voltage, low-amperage jolt of electricity at a distance. (See chapter 9 on Research Associated With the Decision to Use a CED

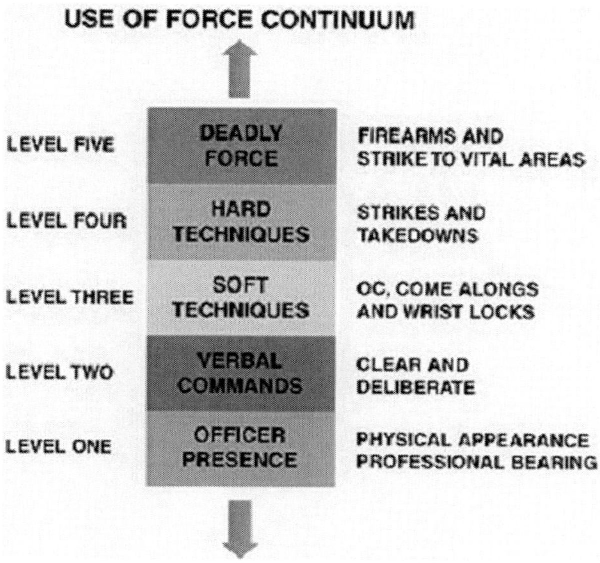

Figure 1. Descriptive diagram of one of many use-of-force continuums.

- **Lethal Force — Officers use lethal weapons to gain control of a situation. Should only be used if a suspect poses a serious threat to the officer or another individual.**
  - Officers use deadly weapons such as firearms to stop an individual's actions.

APPENDIX D. LIST OF ACRONYMS USED IN THIS REPORT

List of Acronyms Used in This Report

CED: Conducted energy device
COD: Cause of death
ECG: Electrocardiograph/electrocardiographic
EDP: Emotionally disturbed person
EMD: Electro muscular disruption
EMS: Emergency medical service(s)
ExD: Excited delirium
JNLWD: Joint Non-Lethal Weapons Directorate
kJ: kilojoule

kV: kilovolt
LED: Light-emitting diode
mA: milliampere
mJ: millijoule
MOD: Manner of death
NIJ: National Institute of Justice
NMI: Neuro muscular incapacitation
PEA: Pulseless electrical activity
USB: Universal service bus
V: volt
VF: Ventricular fibrillation
VT: Ventricular tachycardia

In: Tasers and Conducted Energy Devices ISBN 978-1-62257-463-6
Editors: L. R. Elliott and A. W. Riley

Chapter 3

POLICE USE OF FORCE, TASERS AND OTHER LESS-LETHAL WEAPONS*

United States Department of Justice

ABOUT THIS REPORT

This study looked at injuries that occur to law enforce-ment officers and citizens during use-of-force events. Most applications of force are minimal, with officers using their hands, arms or bodies to push or pull against a suspect to gain control. Officers are also trained to use various other force techniques and weapons to overcome resistance. These include less-lethal weapons such as pepper spray, batons or conducted energy devices (CEDs) such as Tasers. They can also use firearms to defend themselves or others against threats of death or serious bodily injuries.

WHAT DID THE RESEARCHERS FIND?

This study found that when officers used force, injury rates to citizens ranged from 17 to 64 percent, depending on the agency, while officer injury

* This is an edited, reformatted and augmented version of the U.S. Department of Justice, Office of Justice Programs, NCJ 232215, dated May 2011.

rates ranged from 10 to 20 percent. Most injuries involve minor bruises, strains and abrasions.

The study's most significant finding is that, while results were not uniform across all agencies, the use of pepper spray and CEDs can significantly reduce injuries to suspects and the use of CEDs can decrease injuries to officers.

The researchers assert that all injuries must be taken seriously. When police in a democracy use force and injury results, concern about police abuse arises, lawsuits often follow and the reputation of the police is threatened. Injuries also cost money in medical bills for indigent suspects, workers' compensation claims for injured officers or damages paid out in legal settlements or judgments.

WHAT WERE THE STUDY'S LIMITATIONS?

In many cases, agency-supplied injury data did not allow for a detailed analysis of the nature or seriousness of the injuries reported.

INTRODUCTION

Police weaponry has come full circle.

During the middle of the 19th century, police officers in New York and Boston relied on less-lethal weapons, mostly wooden clubs. By late in the century, police departments began issuing firearms to officers in response to better armed criminals. Although firearms are still standard issue, law enforcement agencies are again stressing the use of less-lethal weapons rather than firearms.[1]

The Fourth Amendment forbids unreasonable searches and seizures, and various other legal and policy controls govern how and when officers can use force. Most agencies tightly control the use of force and supervisors or internal affairs units routinely review serious incidents. New technologies have added to the concerns about the use of force by law enforcement.

NEW TECHNOLOGIES RAISE QUESTIONS

During the past 20 years, new technologies have emerged that offer the promise of more effective control over resistive suspects with fewer or less serious injuries. Pepper spray was among the first of these newer less-lethal weapons to achieve widespread adoption by police forces, and more recently, conducted energy devices (CEDs) such as the Taser have become popular.

Taser use has increased in recent years. More than 15,000 law enforcement and military agencies use them. Tasers have caused controversy (as did pepper spray) and have been associated with in-custody deaths and allegations of overuse and intentional abuse. Organizations such as Amnesty International and the American Civil Liberties Union have questioned whether Tasers can be used safely, and what role their use plays in injuries and in-custody deaths.

CEDs such as Tasers produce 50,000 volts of electricity. The electricity stuns and temporarily disables people by causing involuntary muscle contractions. This makes people easier to arrest or subdue. When CEDs cause involuntary muscle contractions, the contractions cause people to fall. Some people have experienced serious head injuries or bone breaks from the falls, and at least six deaths have occurred because of head injuries suffered during falls following CED exposure. More than 200 Americans have died after being shocked by Tasers. Some were normal, healthy adults; others were chemically dependent or had heart disease or mental illness.[2]

Tasers use compressed nitrogen to fire two barbed probes (which are sometimes called darts) at suspects. Electricity travels along thin wires attached to the probes. (A new wireless Taser is also on the market.) Darts may cause puncture wounds or burns. A puncture wound to the eye could cause blindness.[3]

Despite the dangers, most CED shocks produce no serious injuries. A study by Wake Forest University researchers found that 99.7 percent of people who were shocked by CEDs suffered no injuries or minor injuries only. A small number suffered significant and potentially lethal injuries.

This NIJ-sponsored study included six police departments and evaluated the results of 962 “real world” CED uses. Skin punctures from CED probes were common, accounting for 83 percent of mild injuries.[4]

> Several studies found that when agencies adopted the use of pepper spray, they subsequently had large declines in assaults on officers and declines in officer and suspect injury rates, and associated injuries were usually minor. Pepper spray provides a way to reduce injuries.

Policymakers and law enforcement officials want to know whether Tasers are safe and effective, and how (if at all) they should be used to match police use-of-force choices with levels of suspect resistance. This study indicates that CED use actually decreases the likelihood of suspect injury.

PREVIOUS RESEARCH ON USE OF FORCE AND INJURIES

The controversy around Taser use is not unique. Law enforcement agencies found themselves in similar circumstances with pepper spray in the 1990s. Human rights groups such as Amnesty International questioned the safety and misuse of pepper spray as its use spread rapidly in American law enforcement agencies. NIJ funded various studies on the safety and effectiveness of pepper spray.[5]

Some studies have focused on officer injury. Several found that about 10 percent of officers were injured when force was used.[6] However, two studies of major police departments found officer injury rates of 38 and 25 percent.[7] The agencies with lower rates allowed officers to use pepper spray, while the two with higher rates did not.

A few researchers have looked at how various approaches to force affect officer injury rates.[8] Overall, the empirical evidence shows that getting close to suspects to use hands-on tactics increases the likelihood of officer injuries. Research also shows that suspects have a higher likelihood of injury when officers use canines, bodily force or impact weapons such as batons. Alternatives to bodily force and impact weapons are found in other less-lethal weapons such as pepper spray and CEDs.

Previous Studies on Pepper Spray and CEDs

Pepper spray

Law enforcement agencies rapidly adopted pepper spray in the late 1980s and early 1990s as an alternative to traditional chemical agents such as tear gas, but its use sparked controversy. Notably, the American Civil Liberties Union of Southern California asserted that pepper spray was causing in-custody deaths. NIJ studies on the link between pepper spray and in-custody deaths found that the deaths were largely a result of positional asphyxia, pre-existing health conditions or were drug related.[9]

Several studies found that when agencies adopted the use of pepper spray, they subsequently had large declines in assaults on officers and declines in officer and suspect injury rates, and associated injuries were usually minor.[10] Pepper spray provides a way to reduce injuries.

CEDs

Many law enforcement agencies noted that injury rates for officers and suspects declined after they introduced CEDs.[11]

Medical research, including controlled animal trials and controlled human trials, has produced various insights. Some animal studies were conducted to learn if CED use could result in ventricular fibrillation. Several studies showed that standard shocks that lasted five to 15 seconds did not induce ventricular fibrillation of the heart. Higher discharges, 15 to 20 times the standard, or those of longer duration — two 40-second exposures — induced fibrillation or increased heart rhythm in some pigs. In addition, longer exposures led to ventricular fibrillation-induced death in three pigs.[12]

Controlled studies involving healthy human subjects (often law enforcement trainees) found that subjects experienced significant increases in heart rates following exposure, but none experienced ventricular fibrillation.[13]

NIJ Study and Recommendations

NIJ gathered an expert panel of medical professionals to study in-custody deaths related to CEDs. In its report, the panel said that while CED use is not risk free, there is no clear medical evidence that shows a high risk of serious injury or death from the direct effects of CEDs. Field experience with CED use shows that exposure is usually safe. Therefore, law enforcement agencies need not avoid using CEDs provided they are used in line with accepted national guidelines.[14]

A preliminary review of deaths following CED exposure found that many are associated with continuous or repeated shocks. There may be circumstances in which repeated or continuous exposure is required, but law enforcement officers should be aware that the associated risks are unknown. Therefore, caution is urged in using multiple activations.[15]

The seeming safety margins of CED use on normal healthy adults may not be applicable to small children, those with diseased hearts, the elderly, those who are pregnant and other at-risk people. The use of CEDs against these populations (when recognized) should be avoided, but may be necessary if conditions exclude other reasonable choices.[16]

A suspect's underlying medical conditions may be responsible for behavior that leads law enforcement officers to subdue him or her. Sometimes this includes CED use. Abnormal mental status in a combative or resistive subject, sometimes called "excited delirium," may be associated with a risk for sudden death. This should be treated as a medical emergency.[17]

THE NATIONAL SURVEY

The Police Executive Research Forum conducted a survey of state, county and municipal law enforcement agencies to learn more about less-lethal technologies and related policies and training. More than 500 agencies participated.

Most agencies have a "use-of-force continuum" that is covered in training, where officers learn to use suitable force levels depending on circumstances. For example, an officer might start by using verbal commands when dealing with a suspect. Then an officer might move to soft empty-hand tactics (such as pushing) when faced with lack of cooperation or mild resistance. The continuum covers various circumstances up to the use of firearms.

The survey included various levels of resistance and asked agencies to describe what force they allow in each. Most agencies allow only soft tactics against a subject who refuses, without physical force, to comply with commands. Just under half allow officers to use chemical weapons at that point. However, if the subject tensed and pulled when an officer tried to handcuff him or her, most agencies would allow chemical agents and hard empty-hand tactics, such as punching. Many also allow for CED use at this point but about 40 percent do not. Almost three-fourths allow CED use if the suspect flees, and almost all allow it when the subject assumes a boxer's stance. Most agencies do not allow baton use until the subject threatens the officer by assuming the boxer's stance.

Three-fourths of the surveyed agencies that use CEDs issued them between 2004 and 2006. Most are using Tasers. In most agencies, officers receive four or six hours of training, and 63.7 percent of agencies require that officers experience activation (i.e., get shocked) during training.

Most agencies do not allow CED use against a subject who nonviolently refuses to comply with commands. However, six in 10 allow for CED use against a subject who tenses and pulls when the officer tries to handcuff him or her. Agencies usually place the CED with chemical agents in their force continuum, meaning that their use is typically approved in the same circumstances in which pepper spray use is allowed. CEDs are usually lower on the continuum than impact weapons.

> The seeming safety margins of CED use on normal healthy adults may not be applicable to small children, those with diseased hearts, the elderly, those who are pregnant and other at-risk people. The use of CEDs against these populations (when recognized) should be avoided but may be necessary if conditions exclude other reasonable choices.

One facet of the controversy surrounding CED use concerns vulnerable populations and circumstances that pose potentially heightened risk to the subject. For only one circumstance — when a subject is near flammable substances — do most agencies (69.6 percent) ban CED use.

Some 31 percent forbid CED use against clearly pregnant women, 25.9 percent against drivers of moving vehicles, 23.3 percent against handcuffed suspects, 23.2 percent against people in elevated areas and 10 percent against the elderly. However, many agencies, while not forbidding use in these circumstances, do restrict CED use except in necessary, special circumstances.

Analysis of Information from Specific Law Enforcement Agencies

Looking at the experiences of specific agencies can yield important information that might otherwise be lost in larger analyses. The researchers used various statistical techniques to identify factors that increase or decrease the odds of injury to officers and suspects alike.

Richland County Sheriff's Department

The Richland County Sheriff's Department (RCSD) includes about 475 sworn officers who serve the unincorporated portions of Richland County,

S.C. Deputies carry Glock .40 caliber pistols, collapsible metal batons and pepper spray. Increasingly, they also carry the model X-26 Taser. The agency started phasing in Taser use in late 2004. During data collection, about 60 percent of deputies carried Tasers.

Researchers coded 467 useof-force reports covering the period from January 2005 to July 2006. Of the 49 separate injuries recorded for officers (three officers had more than one injury), 46 involved bruises, abrasions or cuts. The department recorded 92 suspect injuries; 69 of those were bruises, abrasions or cuts. Most of the remaining suspect injuries were dog bites, but three involved broken bones or internal injuries.

Further analysis of the data included identifying how various factors increased or decreased the risk of injury to officers or suspects. The use of soft empty-hand techniques by an officer, active aggression by a suspect and suspect use of deadly force all increased the risk for deputies.

Soft empty-hand control was the most frequent force level used by deputies, occurring in 59 percent of all use-of-force incidents. These techniques increased the odds of officer injury by 160 percent. Thus, deputies were at greatest risk for injury when using the least force possible.

Two variables significantly decreased the risk for suspects. Pepper spray use decreased the odds of suspect injury by almost 70 percent, and a deputy aiming a gun at a suspect reduced injury odds by more than 80 percent (because the act of pointing a gun alone often effectively ends the suspect's resistance).

However, the use of a canine posed, by far, the greatest injury risk to suspects, increasing injury odds by almost 40 fold. Suspects who displayed active aggression toward deputies were also more likely to suffer injuries. CED use had no effect on the likelihood of injury; this is inconsistent with the experiences of other agencies, suggesting that not every agency's experience with the Taser will be the same.

Miami-Dade Police Department

The department has about 3,000 officers, is the largest law enforcement agency in the Southeast and is one of the largest departments that has never issued pepper spray to its officers.[18]

The researchers examined 762 use-of-force incidents involving a lone officer and a lone suspect that occurred between January 2002 and May 2006. About 70 percent of the officers carried Tasers by May 2006. Officers were substantially less likely to be injured than suspects, with 16.6 percent (124) of officers injured and 56.3 percent (414) of suspects injured. Most injuries were minor, but 73 suspects (17 percent) suffered serious injuries. Minor injuries included bruises, sprains and lacerations. Major injuries included bites, punctures, broken bones, internal injuries and gunshot wounds.

The department does not issue pepper spray to its line officers, and there were few incidents involving guns or batons. Analysis of the incidents found that the use of both soft-hand tactics and hard-hand tactics by officers more than doubled the odds of officer injury. Conversely, CED use was associated with a 68-percent reduction in the odds of officer injury.

As for suspects, hands-on tactics increased the odds of injury, the use of canines greatly increased the odds and CED use substantially decreased the odds.

Seattle Police Department

The Seattle Police Department has about 1,200 sworn officers. The agency started using Tasers in December 2000. Other less-lethal weapons include pepper spray, batons and shotgun beanbag rounds. The department recorded 676 use-offorce incidents between Dec. 1, 2005, and Oct. 7, 2006. Suspects suffered injuries in 64 percent of the incidents, while officers suffered injuries in 20 percent of the incidents. Officers used hands-on tactics in 76 percent of the incidents. The next most frequent type of force officers used was the Taser (36 percent), followed by pepper spray (8 percent).

Suspects were impaired by alcohol, drugs or mental illness in 76 percent of the incidents. Just over half (52 percent) of the suspects were nonwhite, and 95 percent were male. Analysis of the data revealed that Taser use was associated with a 48-percent decrease in the odds of suspect injury but did not affect officer injury.

The use of unarmed tactics by officers increased the odds of officer injury 258 percent. The odds of officer injury increased significantly when suspects resisted using physical force or the use or threat of use of a weapon.

Although results were not uniform across the agencies, the analysis shows that the use of pepper spray and CEDs can have a significant and positive injury-reduction effect.

Interestingly, nonwhite suspects were less likely to be injured than whites in both agencies (Miami and Seattle) where suspects' race was available as a variable for analysis. Another important finding concerns the use of canines. While canines were used rarely, their use substantially increased the risk of injury to suspects in two of the agencies.

Combined Agency Analysis and Its Limitations

The researchers also conducted a combined analysis of use-of-force data from 12 large local law enforcement agencies.[19] The full report gives a detailed description of the information available and the limits of the data. Most agencies, for example, had details about demographic characteristics of suspects, but only four had officer demographic information. Moreover, the Miami-Dade Police Department did not use pepper spray while San Antonio did not use CEDs.

Despite the limitations, the study's use of a large sample, representing more than 25,000 use-of-force incidents, allowed the researchers to use statistical techniques in an effort to learn which variables are likely to affect injury rates to officers and suspects. The use of physical force (hands, feet, fists) by officers increased the odds of injury to officers and suspects alike. However, pepper spray and CED use decreased the likelihood of suspect injury by 65 and 70 percent respectively. Officer injuries were unaffected by CED use, while the odds of officer injury increased about 21 percent with pepper spray use.

The researchers noted the 12-agency analysis yielded puzzling results about the relationship between pepper spray use and officer injury rates. Those results are inconsistent with the single agency analysis. More research may explain the differences.

Longitudinal Analysis

The researchers reviewed use-of-force information from police departments in Austin, Texas, and Orlando, Fla., to learn how introducing CEDs affected injury rates. This quasi-experimental approach tracked injuries before and after CED introduction.

The Orlando data include 4,222 incidents covering 1998 to 2006. CED use began in February 2003. The Austin data includes 6,596 incidents from 2002 to 2006. However, CED use was phased in beginning in 2003 and was not completed until June 2004. A large drop in injury rates for suspects and officers alike occurred in both cities following CED introduction.

In both cities, Taser adoption was associated with a statistically significant drop in average monthly injuries to suspects. In Orlando, the suspect injury rate dropped by more than 50 percent compared to the pre-Taser injury rate. In

Austin, suspect injury rates were 30 percent lower after full-scale Taser deployment.

In Orlando, the decline in officer injury rates were even greater than for suspects; the average monthly rate dropped by 60 percent after Taser adoption. In Austin, officer injuries dropped by 25 percent.

Interviews with Officers and Suspects

Researchers conducted interviews with 219 officers from South Carolina's Richland County Sheriff's Department, 35 from the Columbia Police Department (CPD), and 35 suspects involved in use-of-force situations to supplement and add a qualitative context to their quantitative analyses. Generally, they tried to contact officers and suspects within 48 hours of receiving a use-of-force report. Interviews were voluntary, and some officers and suspects declined to participate.

In nine out of 105 use-of-force incidents, Richland County Sheriff's Department officers reported that a Taser did not work properly or did not have the desired effect. In addition, researchers received reports of multiple Taser hits on a suspect and multiple uses of the Taser in "drive stun" mode (when the Taser is pressed against a suspect rather than firing darts) to control suspects (or, based on the suspects' reports, as punishment). These reports indicate that some officers are using Tasers multiple times during an encounter.

Nine percent of the officers reported injuries, almost all of which were scrapes, cuts or bruises suffered while struggling with resistant suspects. Officers also reported that 26 suspects (12 percent) were injured. Most suspect injuries were cuts or abrasions, but there were also two dog bites, and one suspect was shot in the arm after firing at officers.

In 22 cases, researchers interviewed both the officers and suspects involved in an incident. Most suspects said officers used excessive or unnecessary force to subdue them. Some suspects said officers used Tasers quickly, and several said the officers enjoyed watching them endure the pain. Some suspects said officers kneed them in the back and kicked or punched them after they were in handcuffs. Some also said officers used Tasers on them after they were handcuffed.

Suspects often tell a different story than the officers who arrest them. In almost all cases, suspects said officers used excessive force and that they were not resisting arrest. The officers, for their part, said they used minimal force to control suspects, and did not mention using force after a suspect was under control. Officers reported that the force used was necessary and reasonable. In a typical account, a suspect said he was unaware there was a warrant out for his arrest, and when police confronted him, he did not resist. He said the officers "pushed me to the ground and put the cuffs on … they didn't have to do that to me." He said that all the officers had to do was tell him to "quit acting up." He complained that officers should just have told him to calm down instead of pushing him to the ground. By contrast, they said the suspect ran away when confronted, so they tackled him. These kinds of contradictions were common; suspects said they did not resist, and officers provided justification for the force levels they used.

In other cases, suspects and officers offered radically different versions of events. For example, in one case, an officer said he saw several traffic violations and the suspects sped off and stopped, with one suspect running away. The officers said the driver then tried to exit the vehicle from the passenger's side holding a shotgun. One officer pointed his weapon at the suspect, who then dropped the shotgun. The suspect failed to mention the shotgun to researchers and only complained that officers put the handcuffs on too tightly and slammed him around in the back of the transport vehicle.

Unlike the Richland County Sheriff's Department, the Columbia Police Department did not use Tasers. The officers described 35 use-of-force incidents. Three officers reported that pepper spray was ineffective. In all three cases, the suspects were either drunk or high on drugs. One case, in particular, highlighted the potential advantages of the Taser over pepper spray in some circumstances. In that case, a 6'7", 370-pound man wanted for domestic violence charged an officer with a metal object in his hand. The officer used pepper spray, but it had no effect. The suspect then retreated to the apartment kitchen and grabbed a knife. The officers pointed their guns at him and ordered him to drop the knife, but he refused. He cut and stabbed himself with the knife while the officers waited for another agency to arrive that was equipped with a Taser. The suspect cut himself more than 100 times before the South Carolina Law Enforcement Division arrived and used a Taser on him. The Taser had an instant effect, and officers were then able to handcuff the suspect.

Most injuries in both agencies occurred when officers and suspects struggled on the ground. The differences between the agencies were striking. RCSD equips most of its deputies with Tasers. The deputies collectively reported fewer injuries to themselves and suspects from ground fighting than did CPD officers. CPD did not issue Tasers, and 31 percent of its officers reported getting cuts, scrapes and bruises from wrestling with suspects on the ground. The prevalence of ground fighting injuries among RCSD officers (less than nine percent) was lower, as were injuries to suspects caused by contact with the ground. Some of the injuries could have been prevented had officers used Tasers instead of hands-on tactics.

Implications for Policy, Training and Future Research

Because of the controversial nature and widespread use of CEDs, the researchers explored their use in detail and made recommendations, based on the findings, for whether and how CEDs should fit into the range of less-lethal force alternatives available to law enforcement officers.

FACTORS AFFECTING INJURIES

Physical Force

The findings clearly show the use of physical force and hands-on control increase the risk of injury to officers and suspects. In Richland County, S.C., soft empty-hand control significantly increased the odds of injury to officers, while hard empty-hand tactics increased the risk of injury to suspects. In Miami-Dade, both types of force increased the risk of injury to both officers and suspects. In Seattle, use of force increased injury risk to officers but not to suspects, while the overall analysis (of 12 agencies) showed increased injury risk to suspects and especially to officers associated with physical force. This increased risk was large. When controlling for the use of CEDs and pepper spray in the overall analysis, using force increased the injury odds to officers by more than 300 percent and to suspects by more than 50 percent.

Suspect Resistance

Increasing levels of suspect resistance were associated with an increased risk of injury to officers and suspects. The increased injury risk was especially acute for officers. In Richland County, active aggression and threats of deadly force increased the odds of officer injury by more than 100 percent. The odds of suspect injury were unchanged in Seattle with increased resistance levels.

These findings suggest that officers, rather than suspects, face the most increased injury risk when suspects resist more vigorously.

Pepper Spray

The findings suggest that, at least for suspects, pepper spray use reduces the likelihood of injury. In Richland County, pepper spray use reduced the odds of suspect injury by 70 percent but did not affect officer injuries. In Seattle, pepper spray use had no effect on injury rates for officers or suspects. However, the overall analysis (of 12 agencies) showed that pepper spray use reduced the likelihood of injury to suspects by 70 percent, which was even more than the decline noted with CEDs (see below). For officers, pepper spray use increased the likelihood of injury by 21 to 39 percent. This finding was unexpected, and more research may help to explain how officers choose to use pepper spray versus CEDs.

CEDs

Except for in Richland County where its effects were insignificant, CED use substantially decreased the likelihood of suspect injury. In Miami-Dade, the odds of a suspect being injured were almost 90 percent lower when a CED was used than when it was not. Similarly, the odds of suspect injury went down by almost 50 percent when CEDs were used in Seattle. The larger analysis of 12 agencies and more than 24,000 use-offorce cases showed the odds of suspect injury decreased by almost 60 percent when a CED was used. In Richland County, Seattle, and in the larger analysis, Taser use had no effect on officer injuries, while in Miami-Dade, officer injuries were less likely when a Taser was used. Controlling for other types of force and resistance, CED use significantly reduced the likelihood of injuries. CED adoption by the Orlando and Austin police departments reduced injuries to suspects and officers over time.

Demographic Characteristics

Apart from officer force and suspect resistance, few other factors influenced injury outcomes. In Miami-Dade, male suspects were twice as likely to be injured as females. The same held true for the 12-agency analysis. In that larger analysis, the presence of a male suspect slightly increased injury risk to officers. In Seattle, female officers were more than twice as likely to be injured as male officers.

PLACEMENT OF PEPPER SPRAY AND CEDS ON A LINEAR USE-OF-FORCE CONTINUUM

People rarely die after being pepper sprayed or shocked with a Taser. However, if injury reduction is the primary goal, agencies that allow use of these less-lethal weapons are clearly at an advantage. Both weapons prevent or minimize the physical struggles that are likely to injure officers and suspects alike. Although both cause pain, they reduce injuries, and according to current medical research, death or serious harm associated with their use is rare. In that sense, both are safe and similarly effective at reducing injuries. Both should be allowed as possible responses to defensive or higher levels of suspect resistance. This recommendation is supported by the findings and is now followed by most agencies that responded to the national survey.

POLICY AND TRAINING ISSUES RELATED TO CEDS

CEDs were used far more often (four to five times more often) than pepper spray among agencies that equipped officers with CEDs and were sometimes used at rates that exceeded empty-hand control. Unlike pepper spray, CEDs do not require decontamination and do not carry the risk of accidental "blow back" that often occurs with pepper spray use. However, they do entail the removal of prongs and the potential for an unintended shock to an officer. Even with these concerns, they are rapidly overtaking other force alternatives. Although the injury findings suggest that substituting CEDs for physical control tactics may be useful, their ease of use and popularity among officers raise the specter of overuse.

The possible overuse of CEDs has several dimensions. CEDs can be used inappropriately at low levels of suspect resistance. Law enforcement executives can manage this problem with policies, training, monitoring and accountability systems that provide clear guidance (and consequences) to officers regarding when and under what circumstances CEDs should be used, or when they should not be used.

Besides setting the resistance threshold appropriately, good policies and training would require that officers evaluate the age, size, gender, apparent physical capabilities and health concerns of a suspect. In addition, policies and training should prohibit CED use in the presence of flammable liquids or in circumstances where falling would pose unreasonable risks to the suspect (in elevated areas, adjacent to traffic, etc.). Policies and training should address the use of CEDs on suspects who are controlled (e.g., handcuffed or otherwise restrained) and should either prohibit such use outright or limit them to clearly defined, aggravated circumstances.

In addition to being used too often, CEDs can be used too much. Deaths associated with CED use often involve multiple Taser activations (more than one Taser at a time) or multiple five-second cycles from a single Taser. CED policies should require officers to assess continued resistance after each standard cycle and should limit use to no more than three standard cycles. Following CED deployment, the suspect should be carefully observed for signs of distress and should be medically evaluated at the earliest opportunity.

DIRECTIONS FOR FUTURE RESEARCH

CEDs can be used too much and too often. A critical research question focuses on the possibility of officers becoming too reliant on CEDs. During interviews with officers and trainers, the researchers heard comments that hinted at a "lazy cop" syndrome. Some officers may turn to a CED too early in an encounter and may relying on a CED rather than rely on the officer's conflict resolution skills or even necessary hands-on applications. Research should explore how officers who have CEDs perceive threats, compared to officers who do not have them. In addition, it is important to determine when, during an encounter, an officer deploys the CED.

Another important CED-related research project would be a case study of in-custody deaths involving CED use and a matched sample of in-custody deaths when no CED use occurred. Advocacy groups argue that CEDs can cause or contribute to suspect deaths. The subjects in CED experimental settings have all been healthy people in relatively good physical condition who are not under the influence of alcohol or drugs. There is no ethical way to expose overweight suspects who have been fighting or using drugs to the effects of CEDs, so an examination of cases where similar subjects lived and died may shed some light on the reasons for the deaths. Law enforcement officials typically argue that most if not all the subjects who died when shocked by a CED would have died if the officers had controlled and arrested them in a more traditional hands-on fight. At this point, the argument is rhetorical and research is needed to understand the differences and similarities in cases where suspects died in police custody, including deaths where a CED may or may not have been involved.

Finally, female officers in Seattle were more than twice as likely to suffer injuries as males. Perhaps the finding in Seattle is an anomaly, but it should be investigated further.

End Notes

[1] Police Executive Research Forum, "Comparing Safety Outcomes in Police Use-of-Force Cases for Law Enforcement Agencies That Have Deployed Conducted Energy Devices and a Matched Comparison Group That Have Not: A Quasi-Experimental Evaluation," report submitted to the National Institute of Justice, grant number 2006-IJCX-0028, 2009: 13.

[2] http://www.ojp.usdoj.gov/nij/topics/technology/less-lethal/how-ceds-work.htm.

[3] Ibid.

[4] http://www.ojp.usdoj.gov/nij/topics/technology/less-lethal/monitoring-ced-use.htm.

[5] Edwards, S.M., J. Granfield, and J. Onnen, *Evaluation of Pepper Spray,* Research in Brief, Washington, D.C.: U.S. Department of Justice, National Institute of Justice, February 1997, NCJ 162358; Granfield, J., J. Onnen, and C.S. Petty, *Pepper Spray and In-Custody Deaths,* Alexandria, Va.: International Association of Chiefs of Police, 1994; Petty, C.S., "Deaths in Police Confrontations When Oleoresin Capsicum Is Used," final report, Washington, D.C.: U.S. Department of Justice, National Institute of Justice, 2004, NCJ 204029.

[6] Henriquez, M., "IACP National Database Project on Police Use of Force," in *Use of Force by Police: Overview of National and Local Data,* Washington, D.C.: U.S. Department of Justice, National Institute of Justice and Bureau of Justice Statistics, 1999: 19-24; Kaminski, R., C. DiGiovanni, and R. Downs, "The Use of Force Between the Police and Persons With Impaired Judgment," *Police Quarterly,* 7 (2004): 311-338; Smith, M.R., and M. Petrocelli, "The Effectiveness of Force Used by Police in Making Arrests," *Police Practice and Research,* 3 (2002): 201-215.

[7] Alpert, G.P., and R.G. Dunham, "Analysis of Police Use-of-Force Data," final report, Washington, D.C.: U.S. Department of Justice, National Institute of Justice, 2000, NCJ 183648; Alpert, G.P., and R.G. Dunham, *Understanding Police Use of Force: Officers, Suspects, and Reciprocity,* Cambridge, NY: Cambridge University Press, 2004; Kaminski, R.J., and D.W.M. Sorensen, "A Multivariate Analysis of Individual, Situational, and Environmental Factors Associated with Police Assault Injuries," *American Journal of Police,* 14 (3/4) (1995): 3-48.

[8] See, e.g., Alpert and Dunham, 2000, note 7.

[9] Granfield, Onnen, and Petty, 1994, note 5; Petty, 2004, note 5.

[10] Edwards, Granfield, and Onnen, 1997, note 5; Kaminski, R.J., S.M. Edwards, and J.W. Johnson, "Assessing the Incapacitative Effects of Pepper Spray During Resistive Encounters With the Police,"
Policing: An International Journal of Police Strategies and Management, 22 (1999): 7-29; Lumb, R.C., and P.C. Friday, "Impact of Pepper Spray Availability on Police Officer Use-

of-Force Decisions," *Policing: An International Journal of Police Strategies and Management,* 20 (1997): 136-148; National Institute of Justice, *The Effectiveness and Safety of Pepper Spray,* Research for Practice, Washington, D.C.: U.S. Department of Justice, National Institute of Justice, 2003, NCJ 195739; Nowicki, E., "Oleoresin Capsicum: A Non-Lethal Force Alternative," *Law Enforcement Technology,* 20 (1993): 24-27; Smith and Petrocelli, 2002, note 6.

[11] Charlotte-Mecklenburg Police Department, *Taser Project: First Year—Full Deployment Study.* Charlotte, N.C.: Charlotte-Mecklenburg Police Department, 2006; Hougland, S., C. Mesloh, and M. Henych, "Use of Force, Civil Litigation, and the Taser," *FBI Law Enforcement Bulletin,* 74 (2005): 24-30; Jenkinson, E., C. Neeson, and A. Bleetman, "The Relative Risk of Police Use-of-Force Options: Evaluating the Potential for Deployment of Electronic Weaponry," *Journal of Clinical Forensic Medicine,* 13 (2005): 229-241.

[12] Dennis, A.J., D.J. Valentino, R.J. Walter, K.K. Nagy, J. Winners, F. Bokhari, D.E. Wiley, K.T. Joseph, and R.R. Roberts, "Acute Effects of TASER X26 Discharges in a Swine Model," *The Journal of Trauma, Injury, Infection and Critical Care,* 63 (2007): 581-590; Esquivel, A., E. Dawe, J. Sala-Mercado, R. Hammond, and C. Bir, "The Physiological Effects of a Conducted Electrical Weapon in Swine," *Annals of Emergency Medicine,* 50 (2007): 576-583; Ho, J.D., J.R. Miner, D.R. Lakireddy, L.L. Bultman, and W.G. Heegaard, "Cardiovascular and Physiologic Effects of Conducted Electrical Weapon Discharge in Resting Adults," *Academic Emergency Medicine,* 13 (2006): 589-595; Lakkireddy, D., D. Wallick, A. Verma, K. Ryschon, W. Kowalewski, O. Wazni, J. Butany, D. Martin, and P.J. Tchou, "Cardiac Effects of Electrical Stun Guns: Does Position of Barbs Contact Make a Difference?" *Pacing and Clinical Electrophysiology,* 31 (2008): 398-408; McDaniel, W.C., R.A. Stratbucker, M. Nerheim, and J.E. Brewer, "Cardiac Safety of Neuromuscular Incapacitating Defensive Devices," *Pacing and Clinical Electrophysiology,* 28 (2005): s284-s287; Nanthakumar, K., I.M. Billingsley, S. Masse, P. Dorian, D. Cameron, V.S. Chauhan, E. Downar, and E. Sevaptsidis, "Cardiac Electrophysiological Consequences of Neuromuscular Incapacitating Device Discharges," *Journal of the American College of Cardiology,* 48 (2006): 798-804; Roy, O.Z., and A.S. Podgorski, "Tests on a Shocking Device — The Stun Gun," *Medical and Biological Engineering and Computing,* 27 (1989): 445-448; Stratbucker, R., R. Roeder, and M. Nerheim, "Cardiac Safety of High Voltage Taser X26 Waveform," Engineering in Medicine and Biology Society, Proceedings of the 25th Annual International Conference of the IEEE EMBS, Cancun, Mexico, 1094-678X, 4 (2003): 3261-3262; Walter, R., A. Dennis, D. Valentina, B. Margeta, K. Nagy, F. Bokhari, D. Wiley, K. Joseph, and R. Roberts, "TASER X26 Discharges in Swine Produce Potentially Fatal Ventricular Arrhythmias," *Academic Emergency Medicine,* 15 (2008): 66-73.

[13] Dawes, D.M., J.D. Ho, M.A. Johnson, E. Lundin, T.A. Janchar, and J.R. Miner, "15-Second Conducted Electrical Weapon Exposure Does Not Cause Core Body Temperature Elevation in Non-Environmentally Stressed Resting Adults," *Forensic Science International,* 176 (2008): 253-257; Dawes, D.M., J.D. Ho, and J.R. Miner, "The Effect of a Cross-Chest Electronic Control Device Exposure on Breathing," *Annals of Emergency Medicine,* 54 (2008): 65; Dawes, D.M., J.D. Ho, M.A. Johnson, E. Lundin, and J.R. Miner, "15-second Conducted Electrical Weapon Application Does Not Impair Basic Respiratory Parameters, Venous Blood Gases, or Blood Chemistries and Does Not Increase Core Body Temperature," *Annals of Emergency Medicine,* 50 (2007): 6; Dawes, D.M., J.D. Ho, M.A. Johnson, E. Lundin, and J.R. Miner, "Breathing Parameters, Venous Blood Gases, and Serum Chemistries With Exposure to a New Wireless Projectile Conducted Electrical Weapon in Human Volunteers," *Annals of Emergency Medicine,* 50 (2007): 133; Ho, J.D., D.M. Dawes, L.L. Bultman, J.L. Thacker, L.D. Skinner, J.M. Bahr, M.A. Johnson, and J.R. Miner, "Respiratory Effect of Prolonged Electrical Weapon Application on Human Volunteers," *Academic Emergency Medicine* 14 (3) (2007): 197-201; Ho, J.D., J.R. Miner, D.R. Lakireddy, L.L. Bultman, and W.G. Heegaard, "Cardiovascular and Physiologic

Effects of Conducted Electrical Weapon Discharge in Resting Adults," *Academic Emergency Medicine,* 13 (2007): 589-595; Ho, J.D., D.M. Dawes, R.F. Reardon, A.L. Lapine, and J.R. Miner, "Echocardiographic Determination of Cardiac Rhythm During Trans-Thoracic Wireless Conducted Electrical Weapon Exposure," *Annals of Emergency Medicine,* 52 (2008): 62; Levine, S.D., C. Sloane, T.C. Chan, J. Dunford, and G. Vilke, "Cardiac Monitoring of Human Subjects Exposed to the Taser," *Journal of Emergency Medicine,* 13 (2007): 47; Levine, S.D., C. Sloane, T.C. Chan, G. Vilke, and J. Dunford, "Cardiac Monitoring of Subjects Exposed to the Taser," *Academic Emergency Medicine,* 12 (2005): 71; Vilke, G.M., C. Sloane, K.D. Bouton, F.W. Kolkhorst, S. Levine, T. Neuman, E. Castillo, and T.C. Chan, "Physiological Effects of a Conducted Electrical Weapon on Human Subjects," *Annals of Emergency Medicine,* 26 (2007): 1-4.

[14] National Institute of Justice, *Study of Deaths Following Electro Muscular Disruption,* Special Report, Washington, D.C.: U.S. Department of Justice, National Institute of Justice, 2011: 3, NCJ 233432.

[15] Ibid., 4.

[16] Ibid.

[17] Ibid., 5.

[18] The MDPD provides police services to the unincorporated areas of Miami-Dade County, Fla., which together contain more than 1 million people in a 1,840 square mile area.

[19] The agencies included police and sheriff's departments in Austin, Texas; Cincinnati, Ohio; Harris County, Texas; Hillsborough County, Fla.; Los Angeles (both the city and the county); Miami-Dade, Fla.; Nashville, Tenn.; Orlando, Fla.; Richland County, S.C.; San Antonio, Texas; and Seattle, Wash.

INDEX

A

B

C

D

E

F

G

M

N

O

P

R

S

T

U

V

W

Y